Twenty-First Century Anxieties

CDE Studies

Edited by
Anette Pankratz

Volume 32

Twenty-First Century Anxieties

Dys/Utopian Spaces and Contexts in
Contemporary British Theatre

Edited by
Merle Tönnies and Eckart Voigts

DE GRUYTER

ZiF

ISBN 978-3-11-152993-6
e-ISBN (PDF) 978-3-11-075825-2
e-ISBN (EPUB) 978-3-11-075836-8
ISSN 2194-9069

Library of Congress Control Number: 2022940355

Bibliographic information published by the Deutsche Nationalbibliothek
The Deutsche Nationalbibliothek lists this publication in the Deutsche Nationalbibliografie; detailed bibliographic data are available on the internet at http://dnb.dnb.de.

© 2024 Walter de Gruyter GmbH, Berlin/Boston
This volume is text- and page-identical with the hardback published in 2022.

www.degruyter.com

Table of Contents

Merle Tönnies and Eckart Voigts
Anger, Anxiety and Hope: The Complicit Realities and Engaged/ing Communities of Contemporary British Dys/Utopian Theatre —— 1

Elaine Aston
"Something's Missing": Feeling the Structures of Project Neoliberal Dystopia —— 11

Nicole Pohl
"To Watch is not Enough": Utopia, Performance, and Hope(lessness) —— 27

Vicky Angelaki
Environment, Virus, Dystopia: Disruptive Spatial Representations —— 43

Paola Botham
Towards a Genealogy of the British Feminist Dystopian Play —— 57

Julia Schneider
Performing Utopia? The Contestation of Dystopian Space in Cecelia Ahern's *Flawed* Series —— 73

Trish Reid
Dystopian Dramaturgies: Living in the Ruins —— 87

Luciana Tamas
***A Description of This World as if It Were a Beautiful Place:* From Avant-Garde Destruction to Dys(u)topias —— 103**

Sebastian Berg
The End of Capitalism and the End of Democracy: Dystopian and Critical Utopian Political Economies in an Age of Austerity —— 117

Dennis Henneböhl
Utopian Past and Dystopian Present? Nostalgia in Brexit Britain —— 133

Anette Pankratz
Civil Wars and Republics in Contemporary (Dystopian) Drama —— 149

Matthias Göhrmann
The Spectre of Utopia/Dystopia: The Representation of Anthropogenic Global Climate Change as Culture-War Issue in Richard Bean's *The Heretic* (2011) —— 165

Leila Michelle Vaziri
"I Am the Abyss into Which People Dread to Fall": Encountering Anxiety in Dystopian Drama —— 185

Peter Paul Schnierer
Visions of Hell in Contemporary British Drama —— 201

Ilka Zänger
"Hiding from the World": Dystopian Subjectivity in Martin Crimp's *In the Republic of Happiness* —— 209

Maria Marcsek-Fuchs
"Let the Doors Be Shut upon"... COVID-19: Relocating the Globe Theatre Stage to the Net —— 225

Notes on Contributors —— 243

Index of Names —— 245

Subject Index —— 249

Merle Tönnies and Eckart Voigts
Anger, Anxiety and Hope: The Complicit Realities and Engaged/ing Communities of Contemporary British Dys/Utopian Theatre

In 2018 and 2019, during the COVID-19 pandemic, the so-called "post-apocalyptic sign-in window meme" circulated on social media platforms (Luti). On a window poster, a bookshop advertised that the post-apocalypse had moved from fiction to current affairs.

Fig. 1: Source: Luti

Indeed, as a most unwelcome case in point, the workshop from which this collection has emerged was derailed by a global pandemic and had to be postponed for a year to March 2021 and then took place in a digital format. The event itself, therefore, became embroiled in the debate on whether stage speculations on future dystopia and utopia had not in the past few years acquired a new urgency and immediacy – so much so that they have become the world's current realities. In response to that, theatre and drama are placed in wider cultural contexts in this collection and read against the background of contemporary social and political developments. The different contributions intersect thematically, and this

https://doi.org/10.1515/9783110758252-001

frequently results in interdisciplinary connections between literary, theatre and cultural studies.

Echoing the overwhelming dystopian realities at our workshop, **Vicky Angelaki** notes in her paper how contemporary theatre has become refracted through the lens of the pandemic. Indeed, just like Angelaki, we will be forced to "read back" earlier dystopian and utopian futures from the vantage points marked by our recent pasts of war and infection. We may ask whether the theatre is not in itself a utopian space of coming together in transient affective community, temporary emotional assembly and fleeting dialogues of feeling. The congregation of theatre thus promises an inherent respite from the threat of complicity with a toxic political landscape in a climate where intensified discourses of fear, instability and insecurity dominate the public spheres.

This question informs, for instance, the contribution by **Elaine Aston**, who begins by making real-life dystopia the present condition for many people – a condition not restricted to neoliberal Britain, but certainly prevalent there. She argues that theatre is a space for building the "unhappy archives" introduced by Sara Ahmed, therefore becoming a site of counter-hegemonic opposition. Aston then proceeds to show how – in detail – recent political theatre in Britain adapts Richard Dyer's formula for the mandatory depiction of a better future in popular entertainment. She demonstrates how Laura Wade's microplay *Britain Isn't Eating* (2014), debbie tucker green's *hang* or Caryl Churchill's *Escaped Alone* (2016) encapsulate hope in records of what is missing in contemporary neoliberal Britain. Estrangement plays a central role in the theatrical strategies of these works – disrupting language use as well as spatial and temporal perception to make audience aware of untenable circumstances to which one nevertheless grows accustomed quite easily in everyday reality.

While historically marginalized in analyses of utopianism and dystopianism – which tended to focus on narrative fiction and film in particular –, the theatre itself may turn into a place of co-creating alternative utopian spaces, as Jill Dolan has argued: "Utopian performatives, in their doings, make palpable an affective vision of how the world might be better." (6) We conceptualize theatre, therefore, as the *theatron*, the Greek "viewing place," where audiences – pointedly understood as groups, as "temporary communities" (Dolan 10) – may catch glimpses of future societies, co-creating a discursive location for enacting and debating imminent hopes and challenges, often couched in the "negative" forms of ("critical," Baccolini/Moylan 2–3) dystopia or post-apocalypse. The spectators' immediate emotional responses are of central importance in this process: "The utopian performative, by its very nature, can't translate into a program for social action, because it's most effective as a *feeling*." (Dolan 19)

Thus, our starting point is that the theatre is an inherently political space and a political art form – it literally takes place within the *polis* (from *politēs* "citizen," from *polis* "city"). The roots of Western theatre in Ancient Greece are inherently linked to creating an alternative spatiality of the City Dionysia of the sixth century BC. From the 2000s onwards, dystopian theatre seems to be a central form that has managed to give political concerns an adequate space again – often in a literal sense, as the use of (fictional and stage) space can play an important role in reaching this aim. The spatial aspects inherent in both dystopia/utopia and the theatre as performative setting therefore merit our special attention. As **Julia Schneider** demonstrates with regard to the very different genre of dystopian Young Adult Fiction, performance can both realise utopian possibilities by contesting dystopian spaces and make utopia graspable by giving it a "real" existence right in front of the audience's eyes. Although she of course deals with fictional performance within a narrative world, the processes she discusses can clearly be transferred to creating spaces of resistance on actual stages in front of real spectators – as quite a few other contributors demonstrate by putting the utopian potential of dystopian spaces centre-stage. Indeed, the intersections between the genres, forms or modes of dystopia and utopia can be taken to run through our volume as a central motif. Paradigmatically, Raymond Williams sees the utopian and the dystopian as constituting two sides of the same coin (as Trish Reid – referring back to Raymond Williams – puts it in her paper in the present volume, 96), or – in the metaphor used in Elaine Aston's contribution – "two interconnecting threads of a double-sided fabric" (11). Taken together, the papers collected here then show which spaces these interconnections open up for actual plays and playwrights grappling with the critical representation of contemporary neoliberal reality and the hope of intervening in it in meaningful ways.

Utopia as a genre is most often associated with prose fiction. Whenever stage utopias are discussed, the argument will emerge that visions of a perfect and idealized world may lack essential ingredients of drama – conflict – and may thus be – in the words of playwright Vinay Patel – "inherently undramatic" (qtd. in Tripney). On the other hand, there have always been utopian plays in Britain – although smaller in number – over the last centuries, from Margaret Cavendish's *The Convent of Pleasure* (1668) to Caryl Churchill's *Light Shining in Buckinghamshire* (1976), Howard Brenton's *Greenland* (1988) and (maybe, but this is open to debate) Ella Hickson's *The Writer* (2018). Recent decades have witnessed a remarkable increase in speculative plays that have expanded from the traditional satires and are now most likely to offer "near-future" (Reid) dystopian scenarios. Unsurprisingly, in past years a notable focus on climate-change theatre and dystopian plays has also appeared in the critical response to these emerging stage engagements. At the same time, Mark Fisher's influential concept of "capitalist realism," among others, has exposed the predicament of the utopian imaginary under neoliberalism, as echoed in Margaret Thatcher's and Angela Merkel's statements presenting the crystallized realities of economics and politics as *alternativlos*.

In view of the dystopian dominance on the stage and the overwhelming sense of crisis and defeat permeating recent political theatre, one overarching topic is the widely disseminated claim made by Fredric Jameson that since the end of the twentieth century Western capitalist societies have been unable to constructively imagine positive social and political developments – "our incapacity to imagine [a better] future" (288–289). Can we diagnose an absence or a revival of utopian spaces in contemporary British theatre, a profound "dystophilia"? One standard response would be that utopianism's lack of conflict makes for poor theatre. Or are there traces of a renaissance of utopia, maybe at least in the form of the "critical" (i.e. ambivalent) dystopias noted by Baccolini and Moylan (3–4)?

Explicitly or implicitly, these issues are taken up in many of the contributions to this volume. **Nicole Pohl's** paper tackles "ineffective" idealist utopia head on. Indubitably, the possibility of a utopian future embedded in a nostalgic conceptualisation of the past (one that never was) has waned. Pohl, however, surveys voices that articulate hope, from Ernst Bloch to Rebecca Solnit, and drawing on Giorgio Agamben, diagnoses the re-emergence of utopia as a form of ethical witnessing. While a utopian future rooted in hope and concordance has been deracinated, we can only witness, wander, search, and practice "infinite listening" as a form of ethical witnessing. In Pohl's view, as in Dolan's, utopian performativity and the performing arts in general can generate utopian desire and hope. She thus highlights the radical potential of utopia on stage, seeing performance as activism and theatre as a laboratory where we experiment with living in hell. In this way, even performance failure can become a way of pointing beyond the performance space towards the spectators and their potential for agency.

The term "drama" derives from the Greek "to do/act" after all, so that recent theatre has attempted to embrace the rallying cries of Solnit and others against despair, even on the edge of doom heralded by the most recent IPCC reports: "A revolution is what we need, and we can begin by imagining and demanding it and doing what we can to try to realize it. Rather than waiting to see what happens, we can be what happens" (Solnit). The *theatron* may provide a space where we can test what it is like to "be what happens." Involving the shared experience of performers and audiences, it seems to posit itself as a performative powerhouse of action in spite of "even the most dystopian theatrical universe"

(Dolan 8). Siân Adiseshiah discussed similar questions at our workshop and will publish her findings in the forthcoming monograph *Utopian Drama: In Search of a Genre* (2022). While most of the 21st-century theatrical dystopias pointedly do not suggest in any way that it might be possible to alter the depicted situation, thereby again differing sharply both from traditional dystopia and from earlier British political drama, we may note with Adiseshiah that there is a slight change in this respect in the more recent theatrical world, which seems to re-engage with hopeful narratives.

Vicky Angelaki, for instance, considers the dystopian legacy of Martin Crimp, discussing his *In the Valley* (2013) and Liz Tomlin's *The Cassandra Commission* (2014) versus the monstrous urban capitalism exposed by Saskia Sassen. Engaging with Mark Robson's recent point that visions of the end of the world imply a nucleus of (at least imaginary) survival, she articulates the seeming paradox that an apocalyptic imagination may be hopeful. Reminiscent of Nicole Pohl's approach outlined above, the theatre audience is at the centre of this argument, as the spatial erosion caused by neoliberal capitalism is exposed and possibly even countered by performance spaces and the audience engagement they can draw forth. Invoking Clare Wallace's observation about the "transitive capacity of the affective" in many recent plays (44), Angelaki concurs with **Paola Botham**'s contribution that this anger – which we might describe as passionately agonistic with Chantal Mouffe (1999) – marks out a utopian core within reigning dystopia. In this way, affective responses to scenarios of destruction can give audiences a sense that it is possible to start anew, primarily locating the utopian impulse in the response rather than the play itself. Botham has explored both utopian and dystopian strands in contemporary British theatre since her dissertation in 2009, where she analysed Churchill's *Far Away* as an exemplary postsocialist dystopian future (243). In her contribution, she updates her view on feminist dystopian and eutopian plays and links Caryl Churchill's first professionally produced stage play, *Owners* (1972), to Lucy Kirkwood's *Tinderbox* (2008). Discussing female complicity within dystopian structures, she notes both allegiances and departures from Churchill, but maintains that there is a clear genealogy of feminist political theatre that pivots around that playwright's work.

Some contributions concentrate closely on the different ways of realising a utopian potential through stage performance and techniques which suggest themselves in recent theatrical practices, and many pick up on the pressing issues of catastrophic climate change and environmental disaster. Indeed, the interventions of Rebecca Solnit, together with the recent prevalence of the impending climate catastrophe in cultural discourses, point to a set of new thematic emphases, not least of all urgent engagements with the impending ecological apocalypse. While Julia Hoydis (340) notes that climate change theatre "has been slow to develop so far, at least in comparison to ecocriticism and criticism of 'cli-fi' prose fiction or film," we can nevertheless record a massive increase in theatrical engagements for instance with issues of sustainability, biodiversity, slow violence, intergenerational justice, science and its dissenters, and the anxieties evoked by them.

Trish Reid's paper discusses some prime examples. Diagnosing a "politics of ruin" with reference to Wendy Brown's work, Reid takes neoliberal capitalism to task for having established a catastrophic "authoritarian neoliberalism" and looks for the critique of capitalist technology as well as rays of hope in Chris Thorpe's *Victory Condition* (2017), and the triple-bill *Interference*. Throughout her contribution, she establishes connections between the anxieties about neoliberalism and the dissatisfaction with realism as a mode of presentation, leading to new theatrical approaches with a distinct utopian potential. In a similar vein, **Luciana Tamas** takes a closer look at innovative aesthetic, formal and discursive strategies of expression of the avant-gardes – subsumed under her term "vocabularies." Her paper provides glimpses into the initial avant-garde impulses anticipated by the Romantics and Symbolists in the nineteenth century which veered towards positions of negativity, (self-)destruction and nihilism, culminating in the First-World War and the violent disruptions of the Dadaists and surrealists. Tamas coins the concept of "dys(u)topia" to highlight that ensconced within these violent and destructive vocabularies we can detect a radically utopian critique of dystopian European culture. On that basis, she the proceeds to map out how the work of Forced Entertainment finds new aesthetic arsenals to re-articulate this alternative position.

In their entirety, these contributions thus foreground the theatre's potential for re-directing its audiences' attention from the present to the future and towards imagining new, innovative forms of interaction and of social organisation. Implicitly, **Sebastian Berg**'s paper on the political left in Britain can function as a contrastive foil here, as he charts two different discursive positions in the political left with regard to capitalism's potential for reform: on the one hand, the more fatalistic dystopian outlook personified by Wolfgang Streeck; on the other, David Coates's critical utopian approach which still sees scope for agency and change. When one compares Berg's analysis with the theatrical approaches examined before, it becomes even clearer just how radical the theatre can be in transcending the dystopian present by envisaging utopian futures. The plays can realize such ideas within the space of the stage and/or in the spectators' thoughts and feelings and thus have the chance to give a very direct, concrete edge to left-wing political agency.

Focusing more closely on Britain itself, **Dennis Henneböhl** highlights the nostalgic prioritization of the past in contemporary British politics and society.

His close readings of a broad range of discursive patterns demonstrate how (more or less irrespective of the concrete topic) the present is represented as a dystopia from which the only viable escape route seems to be turning back to a past constructed as a perfect "Golden Age." In this narrative of decline "utopian" thought cannot develop any innovative, radical potential, which yet again stresses the need for the theatre to explore and realise its own specific dys/utopian possibilities to the full (not only) in Britain. Indeed, as **Anette Pankratz'** paper shows by examining Rory Mullarkey's *The Wolf from the Door* as a case study, 21st-century British drama's typical mix of dystopian and utopian elements can also combine with irony and absurdism to scrutinise current "retrotopic" tendencies directly. Pankratz reads this play and other examples as demonstrations of how the past has lost any genuine meaning in British society although (or because?) it is constantly being invoked as a model for the future. While *The Wolf from the Door* pointedly refuses to take an unequivocal political stance, it becomes clear throughout the analysis how the spatial presence characteristic of the theatre implicitly opposes the nostalgic evasion of the present in British society which is diagnosed by Dennis Henneböhl. From a different angle, **Matthias Göhrmann**'s approach to Richard Bean's *The Heretic* also focuses on the combination of dystopian and utopian patterns with satire and comedy. Göhrmann draws on the concept of the "culture war" to demonstrate how the play criticizes the neoliberal present and the social divisions which run through it, leading to a pervasive rhetoric of polarization. The (pseudo-)utopian happy ending then again leaves the ultimate interpretation (and any impulse for fundamental social and political change) to the audience and their response to the theatrical experience.

The collection closes with a set of contributions about plays where the utopian potential remains more hidden and the focus is on extreme anxiety, as **Leila Michelle Vaziri**'s theorization of the temporality of anxiety and pain makes abundantly clear. Following Sara Ahmed, she describes anxiety as an ever-intensifying conglomeration of several fear-inducing objects which can make it impossible to pin down the exact cause of fear. For Vaziri, dystopian plays use the multimediality of theatre to articulate feelings of anxiety and pain as a linguistic and theatrical foreboding of an imminent breakdown, signalling the dissolution of chronological structure and a mix-up of temporal levels. Discussing works from 2015 (Zinnie Harris's *How to Hold Your Breath*) and 2016 (Alistair McDowall's 2016 play *X*), she diagnoses anxiety in the emerging dystopian alternative realities marked by an emotional overload. We might also be able to apply this feeling of generalized anxiety to **Peter Paul Schnierer**'s discussion of hell as the original dystopia. Schnierer illustrates the difficulty of spatial mapping for a space which has traditionally become a locus of visions and revelations for human anxieties. He asks why contemporary plays resort to established patterns to exorcise contemporary demons – and his analysis of Simon Stephens's *Pornography* (2007) and two plays that premiered in 2015, Alistair McDowall's *Pomona* and Zinnie Harris's *How to Hold Your Breath*, suggests that hell (which is similarly invoked in Pohl's paper) continues to be a valid reference point for new evils precisely to render them aesthetically – and affectively – manageable. Although the term itself does not appear in her paper, **Ilka Zänger**'s reading of selected theatre texts by Martin Crimp can also be taken to examine various kinds of "hell." As she explains, Crimp focuses the construction of his dystopias on the effects which neoliberalism has on individuals. The immediacy of the dystopian tropes, combined with the dissolution of traditional dramatic form, allows the works to communicate the crisis of the individual to the spectators in a very direct, emotional form, constantly heightening their unease and potentially resulting in the kind of anxiety and pain discussed by Vaziri's with regard to *How to Hold Your Breath* and *X*. At least for the duration of the performance, hope has clearly faded into the background or even evaporated completely.

We began this introduction by assessing theatre as a hopeful place that provides a locus for societies to perform their potential dys/utopian futures. After the COVID pandemic, however, theatre itself has become a precarious and potentially endangered site, encroached upon by pressures of economic viability. **Maria Marcsek-Fuchs** intervenes – in conjunction with many other researchers in the past two or three years – to discuss how theatre ought to react to COVID as an institution in order to preserve its a utopian potential, even when real spaces have to be replaced by virtual ones. Marcsek-Fuchs highlights the importance of low thresholds and accessibility to the *theatron* as a utopian space, which may not just be temporarily exchanged for, but also enhanced through online performances and new (digital) performative spaces. Her key question is how to invite wider audience participation in digital formats and how one can combine the utopian digital elements with the physical performances now possible again.

Indeed, the highly productive and inspiring workshop on which this collection is based can also be seen as a case study of a "utopian" community establishing itself in a virtual "performative" space, including both the "performers" and the audience. We are very grateful to *Zentrum für interdisziplinäre Forschung* (Bielefeld) for providing such perfect logistic and technical support for this event, and to all speakers and participants. Special thanks are due to *ZiF* for their financial contribution to the publication of the results and to all our contributors for their papers and their patience throughout the editing process. We would also like to express our gratitude to the student assistants at Paderborn and Braunschweig for the time and trouble they invested in the nitty gritty details of editing: Alexandra Diekhof, Antonie Huff, Amelie Schmedemann – this is your applause!

Works Cited

Adiseshiah, Siân. *Utopian Drama: In Search of a Genre*. London: Bloomsbury, forthcoming 2022.
Baccolini, Raffaela, and Tom Moylan. "Introduction: Dystopia and Histories." *Dark Horizons: Science Fiction and the Dystopian Imagination*. Ed. Raffaela Baccolini and Tom Moylan. New York: Routledge, 2003. 1–12.
Botham, Paola. *Redefining Political Theatre in Post-Cold War Britain (1990–2005): An Analysis of Contemporary British Political Plays*. PhD Coventry University, 2009.
Brown, Wendy. *In the Ruins of Neoliberalism: The Rise of Antidemocratic Politics in the West*. New York: Columbia UP, 2019.
Dolan, Jill. *Utopia in Performance: Finding Hope at the Theater*. Ann Arbor: U of Michigan P, 2005.
Fisher, Mark. *Capitalist Realism: Is There No Alternative?* Winchester: Zero Books, 2009.
Hoydis, Julia. "(In)Attention to Global Drama: Climate Change Plays." *Research Handbook on Communicating Climate Change*. Ed. David C. Holmes and Lucy M. Richardson. Cheltenham: Elgar, 2020, 340–348.
Jameson, Fredric. *Archaeologies of the Future: The Desire Called Utopia and Other Science Fictions*. London: Verso, 2005.
Luti, Greg. "Post-Apocalyptic Sign-In Window: Literary Meme." *Pens and Words*, 15 May 2020. Web. 29 May 2022. <https://www.pensandwords.com/post/post-apocalyptic-sign-in-window-literary-meme>.
Mouffe, Chantal. "Deliberative Democracy or Agonistic Pluralism?" *Social Research* 66.3 (1999): 745–758.
Reid, Trish. "The Dystopian Near-Future in Contemporary British Drama." *Journal of Contemporary Drama in English* 7.1 (2019): 72–88.
Robson, Mark. "Theatre at the End of the World." *Affects in 21st-Century British Theatre: Exploring Feeling on Page and Stage*. Ed. Mireia Aragay, Cristina Delgado-García, and Martin Middeke. London: Palgrave Macmillan, 2021. 239–256.
Solnit, Rebecca, "Don't Despair: The Climate Fight Is only over if You Think It Is." *The Guardian*, 14 Oct. 2018. Web. 29 May 2022. <https://www.theguardian.com/commentisfree/2018/oct/14/climate-change-taking-action-rebecca-solnit>.
Tripney, Natasha. "An Impossible Dream? The Trouble with Utopian Dramas." *The Guardian*, 30 Dec. 2020. Web. 29 May 2022. <https://www.theguardian.com/stage/2020/dec/30/utopia-theatre-jez-butterworth-jerusalem-come-from-away-this-changes-everything>.
Wallace, Clare. "Moving Parts: Emotion, Intention and Ambivalent Attachments." *Affects in 21st-Century British Theatre: Exploring Feeling on Page and Stage*. Ed. Mireia Aragay, Cristina Delgado-García, and Martin Middeke. London: Palgrave Macmillan, 2021. 43–61.

Elaine Aston
"Something's Missing": Feeling the Structures of Project Neoliberal Dystopia

Dystopia. Utopia. Two interconnecting threads of a double-sided fabric. Both are concerned with imagining futures predicated upon present (and past) understandings of how the world is. There are dystopian, nightmare futures to be feared; utopian worlds to be dreamt of and wished for. Go looking for "utopia" and you are more likely to find it in fiction than theatre: in the literary, utopian genre characterised by the representation of a world transformed by socially progressive values. Equally, looking in the other direction, strange encounters with a futuristic dystopia take place more often in literature, film, or television drama. As *Guardian* critic Michael Billington observes, dystopian theatre is "much rarer" than in fiction: "theatre's strength lies less in futuristic visions than fidelity to the here and now."

However, because real-life dystopia is a *present* condition for many people in neoliberal Britain, this explains why, with its "fidelity to the here and now," 21st-century British theatre reflects a prevalence of dystopic themes, concerns, or issues. Further, when theatre engages with the social lacks created by the dystopian inequalities and injustices of the world that is, it has the capacity to elicit utopian yearnings for an alternative world that is not yet but might be. Thus, utopian sensibilities or longings that arise from theatre's treatment of real-life dystopia refer to the feeling, which Ernst Bloch pinpointed by borrowing from Bertolt Brecht, that "something's missing" (qtd. in Bloch 15).

It is beyond the scope of this essay to account for and detail all that lies behind the feeling that "something's missing" in neoliberal Britain. Suffice it briefly, if somewhat reductively, to note that after decades of neoliberal governmentality social inequalities and injustices have escalated, especially in the years of austerity post the banking crisis of 2007–2008. That crisis exposed neoliberalism's economic model as broken. But in the hands of a Conservative, right-wing government – now a majority government post the December 2019 General Election – it keeps staggering on. And the "reason" why the Conservatives "remain the party of rampant inequality, food banks and hedge funds," declares trade unionist and political commentator Andrew Murray, is because "[t]oo many are doing very nicely out of dystopia" (175). That said, "project neoliberal dystopia" has not gone unchallenged. I use the term "project neoliberal dystopia" advisedly: to mark the importance Stuart Hall attaches to the idea that "[n]o project achieves a position of permanent 'hegemony'": "Hegemony has constantly to be 'worked on,' maintained, renewed and revised" (727). As neoliberalism's increasingly economic and morally bankrupt state has shown itself to be decidedly high maintenance, so it has faced counter-hegemonic opposition from disenfranchised citizenries whose demands for the deepening of democracy lie in the undoing of project neoliberal dystopia's hegemony.

During the last ten years or so, this counter-hegemonic opposition has included a significant revival of feminism as a social movement. Furthermore, feminist objections to persistent inequalities in the theatre industry also escalated, notably in the wake of the #Me Too movement. Signs of slow but incremental equality progress in theatre were evinced in more women dramatists writing for the stage, many of whom lent their playwriting voices to critiquing the escalation of social inequalities and injustices, persistent patriarchalism, or widespread ecological destruction (see Aston 1–29). Since women are disproportionately affected by economic maldistribution, or by the erosion of rights and equality, it is not surprising to find women-centred narratives increasing when the spotlight turns on neoliberal dystopia. Three plays by prominent feminist playwrights will enter the frame of this discussion: *Britain Isn't Eating* (2014), an online microplay by Laura Wade; *hang* (2015) by debbie tucker green; and *Escaped Alone* (2016) by Caryl Churchill. Diverse in style, form and content, these works share a critical (in all senses) concern with the idea that "something's missing." Individually, each offers a variation on dystopia/utopia in performance, thereby eliciting different layers of critical reflection; together they attest to and exemplify British playwriting's counter-hegemonic tendency.

British Theatre's "Unhappy Archives"

British theatre's counter-hegemonic disposition is writ large in the annals of playwriting where renderings of a dystopian "here and now" oscillate with a utopian longing for an alternate future. Hence, as it navigates anti-democratic scenes of contemporary life, theatre builds what Sara Ahmed terms "unhappy archives" (18). These are archives oppositional to an unjust socio-economic distribution of happiness; their artefacts scrutinise alleged happiness-making objects for their unhappiness-making properties. Thus, plays in theatre's "unhappy archives" are those that return us time and again to the social displeasures and disparities that prevail, thereby articulating what Ahmed, citing Aldous Huxley's *Brave New World* (1932), terms the "political demand for 'the right to be unhappy'" about the status quo (19). By rehearsing this demand, playwrights feel their way towards "revolutionary forms of political consciousness" that are concerned with "heightening our awareness of *just how much* there is to be unhappy about" 222–223).

With their interest in revealing "*just how much* there is to be unhappy about," theatre's counter-hegemonic "unhappy archives" contrast with entertainment's investment in "happy archives." In his seminal article "Entertainment and Utopia," Richard Dyer pioneered a framework for identifying society's inadequacies and understanding how popular forms of entertainment (musicals) can afford a temporary reprieve from these shortcomings through their rehearsals of utopian solutions. A fundamental function of entertainment, Dyer argues, is to offer an "image of 'something better' to escape into," or to proffer "something we want deeply that our day-to-day lives don't provide" (177). Contrastingly, British theatre, as the notion of "unhappy archives" attests, often entertains the image of something *worse*, something we seemingly cannot escape from, something that deeply and negatively impacts our daily lives.

These contrasting functions appear in Table 1. The left-hand column records Dyer's categories of social inadequacies as originally tabled. The middle column registers the deepening, worsening of those lacks caused by 21st-century project neoliberal dystopia. The third column relists Dyer's utopian solutions – abundance, energy, intensity, transparency, and community – but, except for transparency, as solutions that are withheld, crossed-out. What appears instead is: the abundance of poverty; the animation of neoliberalism as an energy-depleting force; a heightened sense of the non-affectivity of living; the lone individual at risk and community, togetherness threatened. Transparency I retain since what is fundamental to theatre's "unhappy archives" is the opportunity to see the world as it really is – to elicit altered states of perception about what is and what might be. Also, at the bottom of the middle column, I added one more lack to Dyer's original categories: insecurity. This feels a necessary addition given the escalation of globally occurring threats in the 21st century that include (although are not limited to) terrorism, pandemics, and planetary destruction – threats that generate heightened states of terror, anxiety, or fear.

Reflecting on his original table, Dyer observes that his left-hand column evidences a significant omission: "no mention of class, race or patriarchy" (184). The explanation for this exclusion, he elucidates, resides in the way that entertainment is largely (albeit never completely) constrained by capitalism's interests. Contrastingly, some of theatre's "unhappy archives" evidence primary attachments to "class, race or patriarchy," most notably in its specialist collections of feminist artefacts. *Enter Feminist Artefact One:*

Table 1: From Dyer's "Entertainment and Utopia" to Entertaining Project Neoliberal Dystopia in British Theatre

Social tension/inadequacy/absence (DYER)	21st Century Project Neoliberal Dystopia	Entertaining Neoliberal Dystopia in British Theatre
Scarcity (actual poverty in the society; poverty observable in the surrounding societies, e.g. Third World); unequal distribution of wealth	Increased poverty, foodbanks, erosion of state welfare; poverty remains observable in developing regions	~~Abundance~~ ↓ Wealth of economic and social inequalities
Exhaustion (work as a grind, alienated labour, pressures of urban life)	Culture of long hours & doing more with less; increase in work-related stress, depression, and anxiety	~~Energy~~ ↓ Neoliberalism animated as an energy-depleting force
Dreariness (monotony, predictability, instrumentality of the daily round)	Mundane dreariness of life in poverty, daily grind of getting by; lack of time for leisure	~~Intensity~~ ↓ Heightened sense of non-affectivity of living
Manipulation (advertising, bourgeois democracy, sex roles)	Political spin; increased technological control of information; managed imagination ("There is No Alternative," TINA)	~~Transparency~~ ↓ Seeing the world as it really is; altered states of perception
Fragmentation (job mobility, rehousing & development, high-rise flats, legislation against collective action)	Erosion of neighbourhoods (jobs, shops, housing, welfare, community spaces); legislation against trade unionism (diminution of workers' rights; obstacle to collective action)	~~Community~~ ↓ Lone individuals at risk; togetherness under threat
	Insecurity Threats of terrorism, pandemics, planetary destruction	~~Security~~ ↓ Dwelling in fear; heightened anxiety about future

Laura Wade's *Britain Isn't Eating*

Created for an online platform, Wade's *Britain Isn't Eating* was the first in an "Off the Page" series of six microplays devised by the Royal Court Theatre, London, in collaboration with the *Guardian* newspaper in 2014. Launching the series, Wade wanted to explore deepening levels of poverty in neoliberal Britain and the then Coalition government's negative attitude towards the "undeserving" poor. Running at a little over seven minutes, it features a female politician, played by Katherine Parkinson, who has been invited back on to a television show to justify her comments about people on benefits helping themselves to "free food" (Wade). To prove her point, the minister is to deliver a culinary demonstration on how it is possible to make a nutritious meal with meagre, whatever-you-have-left-in-the-cupboard ingredients. All it takes, she insists, is for people to use their "imagination" (Wade). Spinning straw into culinary gold unravels: what to make with a tin of sardines, tomato soup, a few teabags, and a drop of raspberry vinegar? Is the joke on her, she quizzes a TV floor manager? She rallies and announces her intention to conjure a bouillabaisse (posh fish soup!). But the dish is aborted. As the wryly smiling floor manager points out, she must imagine that when your benefits have been sanctioned, you have no money for the electricity meter, no means to cook the food.

The microplay is not funny and not, not funny. Pleasure derives from the satirical unravelling of political spin: the right-wing rhetoric that names the poor as the unworthy poor is exposed for its negative stereotyping of those dependent on state benefits. But dis-easing viewers from laughter generated by ridiculing the political figure is the segue into the representation of the dystopic underside of a nation that houses those who cannot afford to eat. Spatially, juxtaposed with the brightly lit TV-styled kitchen studio is a darkened living room; the "artificiality" (and political artifice) of the former contrasts with the "ambience of a real location" (Bolter): a domestic interior designed to denote the living space of those who go without (sparse furnishings, old-fashioned wallpaper, thin curtains). In a final sequence, a solitary female figure enters the room. She carries a box of basic groceries that mark her out as one of the "undeserving." She never speaks; her facial expressions, appearance and body language denote exhaustion. Mirroring the spatial schism, this disempowered, speech-less woman is also played by Parkinson – a schizophrenic split signifying the divide between the haves and the have-nots, and the power of the one to voice and determine the poverty of the other. The collision and contradiction between the two rooms and bodies is rendered as the politician's voice and image is heard and seen on a small television set in the living room. No longer attempting to conceal her anger, the minister expresses a nightmarish dystopic vision for which, she alleges, the free-loading poor are to blame. To the accompaniment of an eerie, sinister soundtrack, she asks: "How will you feel walking down a high street of boarded up shops knowing that *you* are responsible for their closures?"

(Wade). Thus, blaming the poor for an ailing economy she exonerates neoliberal governmentality for its failure to care.

All told, *Britain Isn't Eating* exemplifies British theatre as the keeper of "unhappy archives" – its commitment to documenting the inequalities of a socially unjust nation. Scarcity, exhaustion, dreariness, manipulation, and fragmentation. All five of Dyer's original social lacks are distilled into the microplay. That it is a microplay and not another *Guardian* article on food poverty is significant. Although creatives were in conversation with experts on food poverty (Cracknell and Gentleman), Wade did not present facts and figures about the rising need for foodbanks, or statistics about those living below the poverty line. As Caryl Churchill once insightfully observed: "A play is a poem not a [political] pamphlet" ("Not Ordinary" 447). Moreover, Churchill elaborated, for a dramatist "to tell us to care isn't enough; we have to be shown what we're to care about in a way that makes us care" ("Not Ordinary" 447). Wade's director, Carrie Cracknell, is of a similar view. She states her interest "in the ways in which stories can open out and shine light on our day-to-day realities and perhaps this way hit you in the heart as well as the mind" (Cracknell and Gentleman). If Wade "makes us care" about food poverty, then this occurs not by "telling us to care," but through affective recognition of the unjust divide between the animated capitalist interests of the one room juxtaposed with the non-affectivity of living in the other. Ultimately, feeling-seeing that divide alters and ironizes the intended meaning of the politician's final line: "We should be cross about this. All of us" (Wade).

Yes, we should be angry about food poverty, but as Wade commented, she finds it "surprising that people aren't more up in arms about inequality," speculating that "[m]aybe it's because everybody is so busy trying to keep their own head above water" (qtd. in Saner). To Wade's observation I would add that this also evinces project neoliberal dystopia's capacity to anaesthetise our power of noticing "something's missing," thereby rendering us immune to *feeling* the structures of inequality. As Ahmed stresses, "feelings of structure" are important because "feelings might be how structures get under our skin" (216). In this regard, what is critical is that "feelings of structure" allow us "to notice what causes hurt" and this involves "unlearning what we have learned not to notice" (Ahmed 216). In short, when even the most devoted readers of the liberal-left *Guardian* may have become immune to the abundance of poverty, Wade's affectively rendered microplay affords a means of noticing "what we have learned not to notice."

The issue of not noticing real-life dystopia also relates to the concern Fredric Jameson expresses about a diminished capacity to imagine "changes in our own society and world" (23). Reflecting on "the waning of the utopian impulse" (23), he insists on the need for "the reawakening of the imagination of possible and alternate futures," for "revolutionary gestures we have lost the habit of performing" (42–43). *Enter Feminist Artefact Two:*

debbie tucker green's *hang*

tucker green is widely acclaimed for her black styling of language and dramaturgy that prioritises the sensory over the semantic. Here again this is not a dramatist "telling" us what to care about. Rather, this is a playwright whose theatre elicits a feeling-structures mode of audience engagement through which harmful structures, the causes of inequality, might get under our skin. *hang*, which premiered at the Royal Court in 2015 under tucker green's direction, is no exception. Set in a time that is specified as "*Nearly now*" (2), the play depicts three characters designated not by names but impersonal numbers: One, Two and Three. One and Two are state officials, representatives of a justice-administering institution. Three is the black, female victim of an undisclosed, horrific crime and she is the one who will now get to decide the method by which the (unseen) perpetrator will die.

As blogger Victoria Sadler commented, Jon Bausor's set design for the meeting room in which the trio assemble had a "near-futuristic dystopian feel" with its black, polished-mirror effect, minimalist furnishings and overhead strip lighting whose automatic sensors Two calls a "bit Star Trek" (3). And a "dystopian near-future" was also how Fiona Mountford described the world of the play "in which all manner of capital punishment is available." And yet, what also needs to be acknowledged is the "utopian impulse" to *hang*. This may not be so easy to recognise. Jameson observes that a "utopian impulse" requires "detective work": a "reading of utopian clues and traces in the landscape of the real; a theorization and interpretation of unconscious utopian investments in realities large or small, which may be far from utopian" (26). Similarly, in the context of theatre such "detective work" involves seeking "utopian clues" in imagined "realities," including those which, as in the case of *hang*, may initially appear to "be far from utopian." Here, my detective work hangs on tracing the play's negative utopian dynamics.

When the play begins the judicial system is subjected to a bitingly comic critique. Despite its futuristic, "bit Star Trek" trappings, the clinically styled office is instantly familiar with its scattering of office chairs and water dispenser that is meant to provide hot and cold alternatives, but always and only offers a lukewarm solution. Familiar too is the Kafkaesque bureaucratic world that One and Two represent – a world of officialdom in which you cannot hold a meeting without having a meeting to discuss said meeting. Transparency in this cog of state machinery is lacking – is a travesty, reduced as it is to protocols and procedures, training manuals and role-playing exercises. What is transparent is the state's failure to offer victim support. With their platitudes and clichés of concern, One and Two cannot come close to imagining the pain, the suffering, that Three and her family have gone through and are left feeling. All told, as Three's pain-fuelled anger collides with the officials' state-scripted, non-affective-care-giving stance, tucker green renders an anti-structural feeling towards a care-less judicial system. And thereby hangs the utopian thread as seminally described by Bloch: the negative utopian that resides in the "critique of what is present" (12), the "'it-should-not-be' of the utopian" (11). It is the critical sensing of "it-should-not-be" that underpins the utopic longing for an alternative political project. Thus, in Three's yearning for transparency, for an end to the setbacks that have kept her waiting three and a half years for justice to be delivered, lies the anti-structural feeling that things could and should be otherwise.

Over the course of the play, this utopian longing is further qualified and identified as the desire to dismantle white, male structures of power. Strategically tucker green withholds one vital clue until close to the end when a reference to the "blue, blue eyes" of the perpetrator identify his race as white (67). This revelation binds the negative utopian thread tightly to the "it-should-not-be" issue of white male supremacy. It makes much sense of Three's determination that the perpetrator will die by hanging: a racial inversion of the white racists' preferred method of punishing, lynching black people. As Michael Pearce attests, this "shock of comprehension" generates a spatial "somewhere in-between" that bridges "the generic British office" with the US and its "American practice of lynching," while temporally blurring "the borders between past injustices and present or future ones" (36–37). That racial injustices haunt this monumental moment of reckoning is also signified in Three's hands that repeatedly tremble. Her hands shake not due to a medical condition but because as she emphatically states and repeats, "this is caused this is *caused*. This was caused" (63). It is caused repeatedly in the present; it was caused in the past.

And yet Three is no down-trodden victim. From the outset she has One and Two on the backfoot, refusing their administrations, upsetting their protocols and significantly derailing proceedings by refusing to allow a letter from the perpetrator – the latest in a long line of developments that she admonishes for their lack of "transparency" – affect her reckoning (28). Her mind is and was firmly made up; he will hang. In the Court's production, Three's command of the space was amplified by Marianne Jean-Baptiste's mesmerising performance in the role as she gave full weight to the beats of Three's rageful silences and long tirade against the irreparably damaged lives of her two young children. All told, as her black woman's anger rails against the injustice of it all, Three/

Jean-Baptiste is the one who tilts the ground on which white supremacy and its idea of "justice" is founded.

This figurative tilting of power creates the kind of anticipatory illumination Bloch posited – a gesture that in its critique of the present, as Burghardt Schmidt explained, anticipates a "realizable future that is reachable no matter how far away" (qtd. in Zipes xxxv). In my introductory remarks I explained that neoliberal dystopia is a project that constantly must be worked on with a determined commitment to proving there can be no alternative, "realizable future." Its systemic modes of power must work in such a way to confine the idea of a not-yet alternative to the realm of the *not ever*. Yet when the lacks (as identified in Dyer's formulation) deepen, so the cracks in the system widen. Potentially this strengthens the negative utopian "it-should-not-be." And as tucker green illuminates past racial injustices, so she anticipates a reckoning that is not not yet, but "nearly now." Indeed, to think of her "nearly now" in the "now" time of writing this essay in 2021, is to think of the Black Lives Matter movement that went global after the killing of George Floyd in the USA in 2020, or of the campaigns around the world to topple statues that memorialise white, male colonialists.

Overall, in its entirety *hang* completes its "revolutionary gesture" by subjugating the white male aggressor to the position of penitentiary supplicant. To conceive the inconceivable erosion of white male supremacy as a possibility is to execute the kind of utopic "operation" Jameson describes – one "calculated to disclose the limits of our own imagination of the future" (23). In the racist imagination of neoliberal dystopia – strengthened by the Brexit debacle – future race relations cannot be imagined except within the borders of white supremacy. To disclose and resist the notion that the future belongs to the racist imagination, tucker green impresses through the negative utopian dynamics of her play. That said, it is important to stress that she does not hold out the false, happiness-making promise of an anti-racist utopian solution. It would be wrong, as Reni Eddo-Lodge observes, to suggest that an "end point" is in sight, that "the post-racist utopia is just around the corner": "Britain's relationship with race and racism isn't a neat narrative with a feel-good resolution" (214). Hence, in its ending, *hang* refuses closure. It ends as Three begins to read the perpetrator's letter; how she will react to his supplication remains an open question, a matter of audience speculation. Not then a closing down but an opening up of what Eddo-Lodge describes as the "uncomfortable" conversations "about racism" that are essential to achieving change which when it happens will be "incremental" and not an immediate, "end point" solution (214).

While neoliberal governmentality has worked hard to maintain the belief that there is no alternative, Jameson observes that the one way we have been able to imagine change is "in the direction of dystopia and catastrophe" (23). *Enter Feminist Artefact Three:*

Caryl Churchill's *Escaped Alone*

Churchill is no stranger to dystopian drama; Billington lists *Far Away* (2000) with its prediction of a totalitarian future in his "top five theatrical dystopias." *Escaped Alone* reprises the scene of global, ecological "dystopia and catastrophe," imagining the world as we know it in the throes of extinction. As a dramatist, Churchill is of the view that in theatre "we do not have to feel, visualize and imagine cautiously," contrasting the production of artistic knowledge with "scientific statements" about the world that are more cautious, limited as they are by the quest for "factual knowledge" ("Not Ordinary" 446). She elaborates: "This doesn't mean making some fanciful definition of life" but creating "a poetic image [which] is a hypothesis which cannot be proved objectively but only by its value and meaning to its writer and audience" (447). If *Escaped Alone*'s dystopian "poetic image" has "meaning and value" it is because we recognise the truth of Churchill's "hypothesis": planetary destruction is the future that awaits if we do not halt project neoliberal dystopia's global exploitation of natural and human resources. In one way this calls to mind Jameson's adage: it is "easier to imagine the end of the world than the end of capitalism" (qtd. in Murray 207). But I also recall the late Sarah Kane saying that "[s]ometimes we have to descend into hell imaginatively in order to avoid going there in reality" (133). Thus, as Churchill plunges us into the "hell" of planetary destruction, she urges the "end of capitalism" as the only means to save the future – a future that is increasingly fragile, damaged, but still to be saved. As Ahmed observes and Churchill imaginatively demonstrates: "we will lose the future if we don't think of the future as something that can be lost" (Ahmed 183).

To conjure the Royal Court's 2016 premiere of *Escaped Alone* directed by James Macdonald, you have to imagine being seated in its downstairs, horseshoe auditorium and faced with a wooden fence stretching the width of the proscenium arch. The fence obscures the view of what lies behind it – a reminder that theatre is a place of looking. In the case of a new play by Churchill, a dramatist from whom we have learnt to expect the unexpected, we anticipate that the act of looking will potentially involve being unseated, dis-comforted, as she invites us not just to look but to look again, especially at "what we have learned not to notice." Our guide to the other side of the fence (which is pulled away), is the ordinary woman Mrs J (Jarrett), played by Churchill veteran Linda Bassett. She takes us into a brightly lit backyard where three older women are seated on an assortment of unmatching chairs. A garden shed, a few plants, a bit of grass. All of this looks familiar, as is the idea of a group of older women whiling away the afternoon. But Mrs J /Bassett who joins the others in idle chit-chat is also the teller of extraordinary tales. She is our guide to the black void which exists in parallel to the garden. Standing on a darkened stage surrounded by two rectangular frames of glowing light, Mrs J/Bassett will deliver seven monologues that report catastrophes of apocalyptic proportions.

Having written about the play several times already, I confess to having something of a compulsive-obsessive relationship with *Escaped Alone*. Most recently I was drawn to analysing Churchill's techniques of *dépaysement* – of "profound disorientation" – deployed to generate altered states of political perception, exploring how these are formed through the women's elliptical conversations; the void which disturbs the familiarity of the garden; and Mrs J's impactful, affect-less style of reporting on disasters conjured in surreal images (Aston 101–103). All these techniques serve to generate "a 'profound disorientation' from the idea that we can go on living as we are when the consequences are world-wide ecological and social destruction" (103).

Given the concerns of this present essay, I also want to single out the uses Churchill makes of security and transparency. The garden is a place in which by rights the women should feel at home, safe and secure. Yet all of them dwell in a place of deep-seated insecurity (the lack I added to Dyer's table), evinced during the course of the play in those moments when each woman breaks the mode of collective chatter to monologue on her respective phobia or anxiety. There is no safeguarding against these individual terrors. The garden in which the women pass the time and time passes cannot be a place of safety. This is because it exists in parallel with the void where natural elements – earth, water, air, and fire – are no longer life-sustaining but life-destroying, courtesy of capitalism's global forces: its "senior executives" (Churchill, *Escaped* 8), the "stock market" (12), or "property developers" (28). Surreal "poetic image" after image conjures this universe as starving, disease-ridden and plagued by a virus (Churchill is nothing if not prophetic). This is not some "far away" planet but the flipside of the happiness-making, wealth-creating, life-sustaining system, capitalism purports to be.

Regarding transparency, already we have seen how project neoliberal dystopia contrives to conceal the harm it causes – the political spin of Wade's politician whose rhetoric demonises the poor, or the state-sanctioned pretence of caring depicted in *hang*. As Hall reflects, "[e]xecutives and corporate spokespersons embroider with impunity, appearing confidently up-beat 'going forwards' even when the economic situation is dire" (722). Contrastingly, Churchill negates this lack of transparency by electing Mrs J the spokeswoman of a world that is not "going forwards" except in the direction of catastrophe. It is her very ordinariness, her lack of social status as a working-class woman that authenticates our sense that she is "telling it like it is" and not "telling it like they [the politicians] want us to hear" (Hall 722). Hence, transparency, acutely lacking in a society enthralled to and configured by capitalism, as Dyer observed, is fulfilled. And yet in fulfilling that want Mrs J does not orientate our vision towards how the world could be better, but rather how it could be worse. If "utopian sensibility" calls "attention to the gap between what is and what could be" (Dyer 185), Churchill's dystopian sensibility narrows the gap between what is and what could be worse. Thus, some consideration of her dominant tendency towards the deepening of dystopian sensibilities is required.

In her seminal *Utopia in Performance*, Jill Dolan posits utopian performatives as those "small profound moments" of intense, shared audience feeling that move and orientate us towards "the possibility of a better future": "Utopian performatives, in their doings make palpable an affective vision of how the world might be better" (5–6). Contrastingly, we might then think of dystopian performatives as performing a reverse operation. No less dependent on moments of intensified affect, dystopian "doings" gesture to how the world might be *worse*. For if a utopian performative "lifts everyone slightly above the present, into a hopeful feeling of what [another] world might be like" (Dolan 5), dystopian performatives sink us into the "present" dystopic condition, thereby generating anxious feelings of the not-yet worse to come. They not only afford a critical sensing of "something's missing" but promise a deeper lack. In Mrs J's "poetic images," as opposed to "scientific statements," we feel-see each diminished, life-sustaining resource further depleted. For instance, she relates how the National Health Service (NHS) "with a three-month waiting time" for "gas masks" (available "privately in a range of colours," she wryly adds) could not deliver life-saving support after "chemicals leaked through cracks in the money" (17). Or, Mrs J reports how "[s]martphones were distributed by charities when rice ran out, so the dying could watch cooking" (22). The NHS in trouble; food programmes screened into homes while people are starving (as Wade's microplay also depicts). Such images oscillate between the deeply dystopian and the absurdly comic. But for all their dystopic futuristic qualities they are immediately recognisable characteristics of the world as we know it; like "utopian sensibility," its dystopian counterpart "has to take off from the real experiences of the audience" (Dyer 185).

What then happens to the hope that Dolan finds in theatre's performance of utopia? "Abandon hope all ye who enter here," as famously inscribed on the gateway to hell in Dante's *Inferno*. Is that what dystopia in performance impresses? The to-be-hoped-for-future that utopia in performance invites its audience to feel-see can never be entirely hope-full since it is coupled with the fear and anxiety of non-wish-fulfilment. Theodor W. Adorno (in conversation with Bloch) clarifies that this is especially so when "the social apparatus has hardened itself against people" (as it has under project neoliberal dystopia), so that the "evident possibility of fulfilment" exists in tension with "the just as evident impossibility of fulfilment" (Adorno in Bloch 4). This being the case, people are then obliged "to identify themselves with this impossibility and to make this impossibility into their own affair" (4). Where utopia in performance grants transitory, fleeting moments of wish fulfilment, dystopia in performance withholds it. But in that withholding lies not the abandonment of hope, but the anxiously hopeful feeling that through the negated "possibility of fulfilment" an audience might come to notice the "social apparatus" and make the alternate future it withholds into "their own affair."

In *Escaped Alone*, after hearing monologue after monologue in which the bodies pile up, there being no antidote to the capitalist plague destroying lives and the planet, we are left with Mrs J's own moment of "terrible rage" (42). "Terrible rage" are the two words she repeats 25 times over. In contrast to the numb, affectless mode of monologic delivery in the void, this last speech by Mrs J/Bassett intensifies in feeling. This is what Ahmed might describe as the "affective knot": the "moment when negative affect spills out" (172). The speech ties together all the dystopian reckonings Mrs J has voiced in the void into this two-word rageful response. Delivered in the garden, its "negative affect" clouds the sunny afternoon that already has grown darker. The rage is hers, but it might be ours too if we feel-see the withholding of an alternate future as our "own affair." In this context, to be rageful, to be animated by negative feelings towards capitalism as an unhappiness-making object, would be hopeful.

Dolan further reflects that "spectators might draw a utopian performative from even the most dystopian theatrical universe" (8). And indeed, Churchill does create one utopian moment that "lifts" the audience "slightly above the present" when the women break off their conversation and burst into a pop song. This is sung not "*to the audience*," but "*for [the women] themselves in the garden*" (28). Suffused in utopian feeling, the musical interlude ruptures the garden conversations in an opposite, happiness-making way to the segues into the void. It amplifies the sense of a women-centred sociability and togetherness as an antidote to the lack of wellbeing. However, although a life-sustaining, reciprocal caregiving force can be found in the company of Churchill's older women (cf. Aston 105), it is rendered vulnerable and eclipsed by dystopic inequalities, injustices and ecological damage that cast a very long shadow. Time is running out, Churchill impresses, for us not not to care about the future.

Final Reflections

It is not in theatre's power to transform society and save the future, but it does have a capacity to strengthen our powers of noticing when it renders "what we're to care about in a way that makes us care." Theatre's acts of noticing depend to a significant degree on techniques of estrangement – on ways of showing us a world that is familiar and yet made strange, so that we are not only invited to look, but to look again. This kind of double-take engagement can take a spatial form, as illustrated in the case of Wade and Churchill's double-world stages (the juxtaposition of the two rooms in the former; the parallel between the garden and the void in the latter). Temporal disturbances might also serve to estrange us from the present, as noted in tucker green's temporal blurring of past, present and future racial injuries in the time of the *"nearly now."* Or we can think again of Churchill's imagining of a dystopian future reported in the past tense by Mrs J whose monologic voicings also temporally disrupt the garden scenes so that the present becomes a dis-continuous present. Language might also be estranged as it is in tucker green's black styling or in Churchill's elliptically formed dialogue and absurdly rendered monologues. Suffused in affect, the sensory, poetic language of these two playwrights elicits a feeling-to-understanding mode of realization that absorbs, takes in, the counter-hegemonic rhythms neoliberalism represses. Indeed, an estranged double-take mode of re-looking depends on affect: we have to be *moved* out of our habit of not seeing, as my analysis of Wade's *Britain Isn't Eating* exemplifies.

For Bloch, the pull towards the utopian through art or literature involved the concept of *Staunen* which, as Jack Zipes explains, can be variously understood as richly encompassing "startlement," "astonishment," "wonder," or "staring" (xxxi). In theatre, to be startled in the way Bloch conceived depends on moments of intensified affect that might potentially generate deeply felt, affective recognition or perception of how the world is, and how it might be. The delayed revelation of the blue eyes of the perpetrator in *hang* is one such moment; Mrs J's "affective knot" of "terrible rage" is another. Ultimately, to be startled into affective recognition of the "it-should-not-be" is a process of dis-illusionment: a way of feeling-seeing through the illusion of neoliberalism as secure in its supposed hegemonic status, thereby illuminating the idea that a not-yet alternative is not only possible but also desirable.

In sum: I have located this trio of feminist artefacts in British theatre's 21st-century "unhappy archives" to explore their counter-hegemonic disposition towards project neoliberal dystopia. All three critique the present dystopic condition, unveiling the social lacks that neoliberalism works hard to conceal. Collectively they remind us that where the neoliberal "social apparatus" conspires to weaken our ability to notice the harm it causes, counter-hegemonic theatre with its affective-poetic image-making is one means by which our capacity to notice might be renewed. As and when the structures of poverty, white male supremacy, or planetary destruction get under our spectators' skin, a critical sensing of how the world is and how it might be otherwise is possible. The theatrical weave of the dystopian/utopian double-sided fabric varies significantly from play to play depending on how dystopic/utopic sensibilities, dynamics, or performatives are threaded. But always it returns us to the negative feeling: "something's missing."

Works Cited

Ahmed, Sara. *The Promise of Happiness*. Durham: Duke UP, 2010.

Aston, Elaine. *Restaging Feminisms*. London: Palgrave Macmillan, 2020.

Billington, Michael. "Never Mind 1984: Michael Billington's Top Five Theatrical Dystopias." *Guardian*, 19 Feb. 2014. Web. 30 Sept. 2021. <https://www.theguardian.com/stage/2014/feb/19/top-five-theatrical-dystopias-1984>.

Bloch, Ernst. *The Utopian Function of Art and Literature*. Trans. Jack Zipes and Frank Mecklenburg, Cambridge: MIT P, 1988.

Bolter, Eben. "Up Close and Political: How We Shot the *Guardian* and Royal Court Microplays." *The Guardian*, 5 Dec. 2014. Web. 30 Sept. 2021. <https://www.theguardian.com/stage/2014/dec/05/guardian-royal-court-microplays>.

Churchill, Caryl. "Not Ordinary, Not Safe: A Direction for Drama?" *The Twentieth Century* (Nov. 1960): 443–451.

Churchill, Caryl. *Escaped Alone*. London: Nick Hern Books, 2016.

Cracknell, Carrie, and Amelia Gentleman. "Food for Thought: Making the *Guardian* and the Royal Court's First Microplay." *The Guardian*, 17 Nov. 2014. Web. 30 Sept. 2021. <https://www.theguardian.com/stage/2014/nov/17/britain-isnt-eating-microplay-off-the-page-royal-court>.

Dolan, Jill. *Utopia in Performance: Finding Hope at the Theater*. Ann Arbor: U of Michigan P, 2005.

Dyer, Richard. "Entertainment and Utopia." *Genre: The Musical*. Ed. Rick Altman, London: Routledge & Kegan Paul, 1981. 175–189.

Eddo-Lodge, Reni. *Why I'm No Longer Talking to White People About Race*. London: Bloomsbury, 2018.

Hall, Stuart. "The Neo-Liberal Revolution." *Cultural Studies* 25.6 (2011): 705–728.

Jameson, Fredric. "Utopia as Method, or the Uses of the Future." *Utopia/Dystopia: Conditions of Historical Possibility*. Ed. Michael D. Gordin, Helen Tilley, and Gyan Prakash, Princeton: Princeton UP, 2010. 21–44.

Kane, Sarah. Interview. *Rage and Reason: Women Playwrights on Playwriting*, Ed. Heidi Stephenson and Natasha Langridge, London: Methuen Drama. 1997. 129–135.

Mountford, Fiona. Review of *hang*. *Evening Standard*, 17 June 2015. Web. 30 Sept. 2021. <https://www.standard.co.uk/culture/theatre/hang-theatre-review-an-hour-that-packs-a-punch-at-the-royal-court-downstairs-10325842.html>.

Murray, Andrew. *The Fall and Rise of the British Left*. London: Verso, 2019.

Pearce, Michael. "Black Rage: Diasporic Empathy and Ritual in debbie tucker green's *Hang*." *debbie tucker green: Critical Perspectives*. Ed. Siân Adiseshiah and Jacqueline Bolton. London: Palgrave Macmillan, 2020. 23–44.

Sadler, Victoria. Review of *hang*. *Huffington Post*, 18 June 2015. Web. 30 Sept. 2021. <https://www.huffingtonpost.co.uk/victoria-sadler/hang-royal-court-theatre_b_7606006.html>.

Saner, Emine. Interview Laura Wade. *The Guardian*, 21 Sept. 2014. Web. 30 Sept. 2021. <https://www.theguardian.com/stage/2014/sep/21/laura-wade-the-riot-club-posh-interview>.

tucker green, debbie. *hang*. London: Nick Hern Books, 2015.

Wade, Laura. *Britain Isn't Eating*. *The Guardian*, 17 Nov. 2014. Web. 30 Sept. 2021. <https://www.theguardian.com/stage/video/2014/nov/17/britain-isnt-eating-microplay-guardian-royal-court-video>.

Zipes, Jack. Introduction. Ernst Bloch. *The Utopian Function of Art and Literature*, 1988. xi–xliii.

Nicole Pohl
"To Watch is not Enough": Utopia, Performance, and Hope(lessness)

In 1965, a time marred by end time events such as Auschwitz, Hiroshima and Nagasaki, Frank Kermode reflected on time and eschatological thinking in literature in his lecture series "The Long Perspectives." In these, he suggested that all literature is in some sense apocalyptical, grafting order onto our essentially human experiences of death but also onto destruction and extinction. He argued that *eschaton* scenarios that can be found in different cultures and religions across the world have offered model paradigms to "make tolerable one's moment between beginning and end" by promising cosmologies of transformation and redemption (2). These paradigms have continued in what Kermode calls "concord fictions," attempting to "unite beginning and end and endow the interval between them with meaning" (190). Those writers writing from and about "the middest" then are required to negotiate the immanence and imminence of the impending apocalypse: "And although for us the End has perhaps lost its naive imminence, its shadow still lies on the crises of our fictions; we may speak of it as immanent" (6). Thus, according to Kermode, apocalyptic fictions create meaning, create new myths about life "in the middest" without necessarily promising redemption, transformation or survival.

The Utopian scholar Fredric Jameson has developed Kermode's *eschaton* into a space that could be filled by utopian desire. "[T]he end of the world," Jameson argues, "may simply be the cover for a very different and more properly Utopian wish-fulfilment" – thus, the apocalypse can hold a seed of hope (*Archaeologies* 199). In this sense, concord fictions are valuable – despite the immanent and imminent threat of apocalypse. Life "in the middest" becomes important to anticipate, to imagine and to possibly help to create another world, a religious or secular Blochean "Not-Yet." This Utopian desire ensures that being "in the middest" is not living death but being imbued with life, with knowledge and experience, and the promise of redemption.

We are again in an apocalyptic crisis or, shall I say, apocalyptic crises and states of emergency – we seemed to have arrived at the terminus of "time's arrow," the end time of human life as we know it, and this time, time might

---

Note: I thank Dr Carina Bartleet, whose generous reading and expertise helped me to formulate my argument about utopia and performance.

https://doi.org/10.1515/9783110758252-003

not discredit the apocalyptic projection nor fulfil the promise of redemption (Gould).

I am writing this chapter in the month where the 6th UN Climate Change Conference of the Parties (COP26) in Glasgow, Scotland, is about to take place. I am also writing this chapter after about one and a half years of the COVID-19 pandemic which had disastrous and manifold consequences on humans and non-human species across the world.[1] COVID-19 has persisted in its many mutations, we have not eradicated it. Its outbreak did not come as a surprise, scientists have warned of zoonotic epidemics for a long time, indicating the clear relationship between climate destruction and pandemics. What is evident is, and COVID-19 has accentuated this, the environmental fracture point is not necessarily one. The pressure points are oil, water, political, agriculture, the world finance system, the biosphere, or spirituality – in short, the system failure of the Capitalocene (Patel and Moore). The real virus is ultimately capitalism and its appendages of imperialism, social injustice (particularly evident with the issue of access to the vaccine, and the mortality rates of BAME citizens of COVID-19), climate destruction, climate injustice, and reactionary ecology.

What is interesting is that the climate emergency only scares a few, but COVID-19 scared many and encouraged people to make sacrifices. As much as we are now in an endemic pandemic (the new normal/post-normal), we are experiencing a chronic, if not escalating climate emergency. As Andreas Malm argues, they are not two different crises, they are in fact "interlaced aspects of what is now one chronic emergency," and our response should target both in unison (91). Malm and others have identified the root cause of the chronic emergency – capitalism, and a "crisis of social reproduction" (Out of the Woods, "Uses of Disaster"). So, the focus on radical change if at all possible is to target the socio-economic and political systems we live in with "climate honesty" as the Green House Think Thank recently demanded (Foster). Transformative adaptation (Foster et al.), deep adaptation (Bendell), ecological Leninism (Malm), or disaster communism (Out of the Woods) all share a common denominator – radical change brought about by radical activism. Malm argues that these "rescue operations" are not utopian, "[s]aving people from asphyxiating in a coal shaft is not to open the door to *Schlaraffenland*," but urgent, pragmatic and radical system changes (163).

---

[1] I am proofreading this chapter knowing that COP26 fell well short of delivering the international commitments that would limit warming globally to 1.5 °C. I am proofreading this chapter in the midst of the invasion of the Ukraine by Russia that highlights not only the real dangers of expansionism and revisionist historiography but also the politics of gas and oil imperialism.

In the following, I will reassess this juxtaposition between the effective "rescue operations" and the seemingly ineffective, idealistic utopia. I propose that responses to the chronic climate emergency and the Anthropocene, particularly in the performance arts, can be and are in fact utopian, based not on the too common misrepresentation of utopia as a perfect state and world (*Schlaraffenland*), but utopia as a form of ethical witnessing (Haas). Ethical witnessing not only gives meaning to our lives, even in the face of extinction and emergency, it makes "tolerable one's moment between beginning and end" (Kermode 2). Ethical witnessing can exist even in hope(lessness), as it creates meaning, a sense of responsibility, agency, and potentiality, even if it is for a posthuman world.

I will draw on a range of different examples to illustrate the possibility of ethical witnessing and utopian performativity in performance art. The ecological turn in performance and theatre studies has not only addressed content, but also the ecologies of the theatre/performance as such – ecodramaturgy thus not only underscores the necessity for theatre to address themes of ecology, extinction, and the Anthropocene, it demands at the same time, ecological production and performance practices (see Woynarski, *Ecodramaturgies*; Bartleet). Performance art, however, can take these concerns further, to echo Marina Abramović, by tapping into / disrupting / challenging the daily lives, rituals, habits and relationships of human societies with a sense of imminent and immanent urgency, and activism, by creating "true reality" – even if this means extinction, crises and disaster (Abramović). I am particularly inspired by the evocative variety of forest dramaturgy (Dee Heddon) in Joseph Beuys' *7000 Eichen – Stadtverwaldung statt Stadtverwaltung* (*7000 Oaks – City Forestation Instead of City Administration*, 1982), *The Forest* (1988) by Richard Wilson, the Theater des Anthropozän / Theatre of the Anthropocene (2019), *The Walking Library* (Dee Heddon, 2012 – ongoing), Lisa Woynarski's *The Celebrated Trees of Nashville, Tennessee* (2012), and Extinction Rebellion's "die-ins." I argue that performance art can be and perhaps should be both ethical witnessing and utopian performativity. In that sense, there is an overlap between protest and theatricality, between performance and pre-figurative pragmatism – illustrated so well in the dialectic of "performance as/is politics" (Rai et al. 1–26).[2]

* * *

---

[2] I agree with Love that we should not, however, fall into the trap of "the hubristic and misguided thinking that would purport to save the planet through performance. Theatre is best seen as a small part of a much larger constellation of climate activism, and we ask too much if we demand

Meaning and hope are two different things. In *Hunting for Hope*, Scott R. Sanders writes that "[t]he first condition of hope is to believe that you will have a future; the second is to believe that there will be a decent world in which to live it" (21). And we are not sure about either at this point in humankind's history.

So what does this mean for utopias and utopian thought? How do we generate and maintain hope that will fuel radical change, will allow us to hope for a future with a decent society? Is this at all possible, feasible or necessary? Do we need hope as part of utopian thinking? In his piece on COVID-19, Agamben rightly reflects, "'There is no sense in anything I do, if the house burns down.'"

You could argue that in the greater scheme of things, we do not matter that much. If we take the viewpoint of Deep Time to qualify the importance of humans in Earth's history, the Anthropocene becomes a mere a boundary event (Haraway, "Anthropocene" 160). Indeed, as Katie Mack explored in *The End of Everything (Astrophysically Speaking)*, our universe will be obliterated eventually – humans and non-human species will adapt up to a point but then, everything will be gone. The cosmic end times will bring no day of judgment, no redemption. Climate destruction will just accelerate this process.

So, if and how can utopias, utopian thought be relevant, even useful? How can the performing arts generate utopian desire, build new and different societies from the ruins around us, give meaning to life "in the middest" through ethical agency, or dare I say it, give us hope?

Jameson's essential argument is that the apocalypse "can hold a seed of hope," and is principally based on Ernst Bloch's understanding of hope as manifold forms of resistance to the status quo. Yet, Bloch distinguishes between "abstract" and "concrete" utopias; the latter is "an anticipatory kind which by no means coincides with abstract utopia dreaminess, nor is directed by the immaturity of merely abstract utopian socialism" (146). Concrete utopia, in opposition to "real" utopianism or "pragmatic" utopianism is critical, it transforms hope into a utopian desire that in turn drives political change (see Wright). But what is really possible in our end times?

The essayist Rebecca Solnit observed seeds of utopian hope in local communities' responses to Hurricane Katrina or the San Francisco earthquake of 1906. According to Solnit, these are "disaster utopias" – immediate and temporary social experiments based on the collective empowerment and resilience (often in the absence of governmental aid or response). Solnit reminds us that

---

that any single production should upend audience members' attitudes and shift the public conversation" (235).

these disaster utopias are based on human resilience, (deep) adaptation, and cooperation that manage to rebuild viable communities on the rubble of the old ones. Solnit references C.E. Fritz, who co-directed the Disaster Research Project in the 1950s. Fritz acknowledged the psychosocial effects of disasters which, contrary to his initial premises, can in fact empower victims, create collective resilience and co-operation. He wrote that

> [t]he widespread sharing of danger, loss, and deprivation produces an intimate, primarily group solidarity among the survivors, which overcomes social isolation, provides a channel for intimate communication and expression, and provides a major source of physical and emotional support and reassurance (63).

What Solnit records are microcosms of "lived utopianism," "living out some portion of a transformed future in the here and now" in different ways (Robertson 251). But these are momentary responses to disasters – what we are facing today is a chronic emergency. This is perhaps why a more radical transformation, based on Solnit's disaster utopia and underpinned by anarchist prefigurative utopianism, is Out of the Woods' appeal for "disaster communism." Whilst I would query the term "disaster" in this context, the collective argues that hope in disaster utopianism is crucial but "building paradise in hell is not enough: we must work against hell and go beyond it" (Out of the Woods, "The Uses of Disaster"). Community involvement, resilience and aid during a disaster is not transformatory in essence as it is momentary and limited in systemic outlook. The collective argue that the world is in a state of emergency and that needs emergency measures – an argument that echoes Andreas Malm whose rhetoric invokes warfare in his plea for "war communism." Out of the Woods acknowledge that "disaster communities, then, are glimpses of hope: microcosms of a world formed otherwise. [...] This hope is vital, yet all too often hope kills us." ("The Uses of Disaster")

The problem then rests in the conceptual framing of the state of the world (crisis, emergency, catastrophe, apocalypse); the paradox of the human condition lies in the principle of hope:

> Hope can be intoxicating in its capacity to disorient and to open a world beyond the given, beyond who "we" are. Or, hope can be toxic, precluding us from acting and living in the now, directing our attention to an imagined future that will render the present (inauthentically) tolerable (Colebrook 324).

If hope can become toxic, so can utopian desire and utopian thought; "utopia [also] has its limits: utopia can be toxic," argues China Miéville:

> Utopias are necessary. But not only are they insufficient: they can, in some iterations, be part of the ideology of the system, the bad totality that organises us, warms the skies, and condemns millions to peonage on garbage screen.

In this train of thought, the radical potential of utopia is at best reduced to a mere means of socio-political critique that can be answered with concrete political reforms, at worst abused as political spin and ideology. Utopia is demoted to a distraction, creating merely "a supposedly happy, harmonious, and non-conflictual space" that serves "to soothe and mollify and to entertain" us, in fact distract us from possible climate collapse and extinction? (Harvey 166–167)

Is therefore utopian desire futile and in fact, illusory? The stark juxtaposition between utopia, pragmatism and realism, between hope and hopelessness somewhat obscures that the problem lies elsewhere, as Rudger Bregmann rightly observes,

> But the real crisis of our times, of my generation, is not that we don't have it good, or even that we might be worse off later on. No, the real crisis is that we can't come up with anything better (20).

Bregmann identifies our lack of vision and imagination; we cannot come up with anything better – not for ourselves but for the planet – human and non-human species alike. Popular narratives of extinction, emergency and disaster have become "the fetishisation of capitalist abstractions" that confine human imagination and agency once again (Tomšič 274, 278). Certainly, as Jameson once declared, "it is easier to imagine the end of the world than to imagine the end of capitalism" (*Seeds of Time* xii).

Hypercritique, "negentropic" poetics, and, as I would argue, ethical witnessing are ways out of this intellectual, ideological and imaginative *impasse* (Stiegler 57). New radical knowledge systems from below and from the human/non-human hybrid need to be developed that underpin radical system change. Malm's evocation of war communism as an answer to the chronic climate emergency is, indeed, such a critical response but it is also one of intellectual provocation:

> So is bringing in Lenin or speaking of ward communism: they would never be needed if it weren't for them serving as indexes of the gravity of our ordeal. Precisely in their remoteness from any currently discernible trajectory, not to say their farfetchedness, lies their truth content. An age this bad can only be reflected in extreme contrasting images. Likewise, every concrete measure proposed here and by many others for coming to terms with nature and ending the "boundless imperialism" against it may well be brushed aside as utopian. They are exactly as utopian as survival (174).

This provocation exemplifies the practice of hypercritique, staging the tensions between disruption, global collapse and extinction. Hypercritique is located "in the material contingencies of expression and language," the evocation of "extreme contrasting images," the conflation of utopia and dystopia (Robinson), "negentropic" imagination (Stiegler) and dreams (Sledmere) – "staying with the trouble" (Haraway).

What do hypercritical poetics look like? They are, as the poet and critic Maria Sledmere suggests, "like a love unrequited, inclined towards death, yet to live" (67). As a critical inquiry and creative practice, hypercritical poetics are marked by grief, death, obligation, responsibility and ultimately, honesty and truth. "Even in the burning house," Giorgio Agamben insists, "language remains":

> A poem written in the burning house is truer, more right, because no one can hear it, because nothing ensures that it can escape the flames. But if, by chance, it finds a reader, then that reader will in no way be able to draw back from the apostrophe that calls out from that helpless, inexplicable, faint clamor. Only someone who is unlikely ever to be heard can tell the truth, only someone who speaks from within a house that the flames are relentlessly consuming.

Agamben's convergence of poetry and philosophy, is, as I understand it, a form of ethical witnessing. The ethical element holds utopian desire, the witnessing reveals the dystopian components of a world in which we witness many conflicting dystopian scenarios. Ethical witnessing can take place without hope, in the face of extinction. Indeed, hope is not the opposite of but in a dialectical relationship with hopelessness – it fuels, as Benjamin Haas suggests, a "revolutionary/critical hope(lessness)," it fills life "in the middest" with meaning in the face of extinction and the apocalypse (292).

* * *

If we pursue this train of thought, utopian thought is embedded in revolutionary and critical hope(lessness), and indeed, the dialectic between hope and hopelessness must be experienced, felt and sensed. Benjamin Haas acknowledges the necessity and impact of utopian desire in performance which as a form of embodied hypercritique could provide a new and transformative method of inquiry. Art and performance, he argues, "articulate both a felt (past/present-oriented) and speculative (future-oriented) attachment, especially with investments toward everyday lived experience, speculative fiction genres (think sci-fi), and activist pursuits" (285). Haas references José Esteban Muñoz here, and the quotidian impulse of utopia. In opposition to disaster utopianism which practises "building paradise in hell," the visual, literary and performance arts experiment with living in hell. The utopian prefiguration here is inverted, to experience and actively witness "living in a damaged world" (Haas 285).

> Witnessing may be all we have left. As a practice, it is the first step toward an ethical relationship to extinction and hopelessness. We must begin to witness the pain we cause others. We must extend the category of others to include all nonhuman animals. To watch is not enough (293).

For Haas then, the revolutionary/critical hope(less) performance moves from being an aesthetic event, an art practice to a practice of radical activism.

How would such radical performance activism look like? Hope(lessness) is a negative yet energetic force, not nihilistic but open to a hypercritical conceptualisation of hope. Ethical witnessing is feeling, and experiencing the weight of death, extinction and hopelessness to activate "an anger that can ensure our extinction does not lead to the complete elimination of life on our planet" (Haas 295). In short, Haas' essay embodies hypercritical ethical witnessing as a call to (performance) activism.

Performance scholars and practitioners may insist, of course, that all theatre and performance is political, taking a stance on the world in its entirety. This avowal is rooted in Berthold Brecht's "Episches Theater," and in Augusto Boal's "Theatre of the Oppressed." An important and contemporary example for our context is Extinction Rebellion's performative acts of resistance which embody, visualise, and choreograph their political concerns through "disruptive aesthetics," in some ways a manifestation of the "postdramatic theatre" (Markussen and Lehmann). Particularly, the Red Brigade "die-ins" are first and foremost performance theatre, and public art:

> They resemble human statues, forming a tableaux silent witnesses [sic], standing to take notice, providing a unifying message. They appear gentle, arms outstretched and often interlocked. The group say the red signifies blood that we all share across species. The white faces also create unity and allow for expressions to be more pointed. Their goal is to evoke empathy through movement and expression, much like the tradition of Butoh dancers. These dramatic flares of red resemble interlocking chains and demonstrate the interconnectivity of all beings and the measured steps needed to address climate change, and each body carries a story, which allows a certain vulnerability. Their slowness and deliberation is precisely the cutting intervention needed. Slow down. Slow down. Learn. Subvert compassion fatigue or even empathy fatigue to give way to the accessible, meaningful, new visions that art can provide. They are thoroughly planned, organized, and choreographed with military precision (Coombs 128–129).

These performances are spectacles of ethical witnessing – they perform our being in and with the world. These performances are not prefigurative but they are underpinned by "ethics of possibility," in which "thinking, feeling, and acting increase the horizon of hope, that expand the field of the imagination, that produce greater equity in what I [Appadurai] have called the capacity to aspire, and that widen the field of informed, creative, and critical citizenship" (Appadurai qtd. in Coombs 134).

Indeed, "utopian performativity" is, as Muñoz understands it, a "manifestation of a 'doing' that is on the horizon, a mode of possibility. Performance, seen as utopian performativity, is imbued with a sense of potentiality" (99).[3] This potentiality is imbued with hope(lessness), with terror and despair yet also some hope of transformation even if only for a post-human world. Perhaps therefore, these performances could also be called interventions. Ultimately, they must fail, as Kenn Watt rightly points out, performance failure does not indicate "a lack of success, but a conscious strategy on the part of the artist to gesture towards outside the performance space" – so the experiences will mobilise the participants/spectators.

* * *

If performance art as/is radical activism and utopian intervention that encourages the audience to experience, think and feel the climate crisis, and confront it with hope, anger, and utopian desire, what about spaces in the theatre space itself? Ecodramaturgy acknowledges the environmental impact of theatre productions and business structures, on the one hand. On the other, it reassesses how theatre can put ecology at its dramatic and conceptual centre. Forest dramaturgy explores the relationship between performance and forests, between the space of performance art (physical, mental, spiritual) and arboreal ecologies (Heddon). Forest dramaturgy recognises the entangled ecologies of human and sylvan life forms in forests and woodlands, and tries to capture, illuminate these through performance and texts. Dee Heddon, who coined the term, curated the *Walking Library*, where reading and walking become "an immersive and moving space, a kind of mobile laboratory" in woodlands and forests *(Walking Library Project)*.

Robert Wilson's multi-media spectacle *The Forest*, premiered at the Theater der Freien Volksbühne, Berlin (1988), adopts a more Deep Time / Forest Time perspective and offers ethical witnessing on the decline and corruption of human civilization. Wilson collaborated with David Byrne, Heiner Müller and Darryl Pinckney in an adaptation of the *Gilgamesh* to the nineteenth-century

---

[3] Jill Dolan calls this "utopian performatives" – a moment where the audiences/spectators/participants are elevated "slightly above the present, into a hopeful feeling of what the world might be like if every moment of our lives were as emotionally voluminous, generous, aesthetically striking, and intersubjectively intense" (5).

era of industrialisation with further texts by Charles Darwin, Jean-Henri Fabre, Edgar Allan Poe, William Wordsworth, and George Lichtenberg. Marranca calls *The Forest* "archival;" for

> *The Forest* he [Wilson] and his collaborators have compiled a new encyclopaedia of texts and images that tells the story of the earth and its peoples, and those beyond the earth, an encyclopaedia made up of stories written in the past and rewritten in the present, images inscribed in the monuments of world cultures in other times and places that also tell of our time and our place (71).

The Forest shifts the focus of human history as success and progress to the history of civilisation as decline and fall. The lens remains anthropocentric; the forest remains a mere mythical place, a Forest of Arden, a *theatrum mundi* that merges "cosmic time with theatrical time" to behold the fraught relationship between humans and nature (70).

The Theater des Anthropozän / Theatre of the Anthropocene, founded in 2019 in conjunction with the Humboldt-Universität zu Berlin, also employed a multi-media, multi-platform approach to explore on stage, in performances, and interventions the conflict between human and non-humans, between the man-made world and nature. *Requiem for a Forest: A Journey into the Forests of Prehistoric Days, Present and Future* is a digital staged reading that mixes myth, poetry and scientific reports on the sylvan ecosystem. Similarly to Wilson, the Deep Time perspective of the *Gilgamesh*, *Edda* and other texts read in the spectacle, shifts the focus from humans to the non-human world. In opposition to *The Forest*, though, the forest becomes the actor in the *Requiem* – a shift that speaks to the premise of eco-dramaturgy to interrogate the anthropocentric world view that underpins not only our understanding of the climate crisis but also our reaction to it. *Requiem for a Forest* is set up as a virtual labyrinth of texts, music, readings – an arboreal thicket of sensual and creative experiences. Frank Raddatz, the founder and creative director of the Theater des Anthropozän / Theatre of the Anthropocene writes:

> Instead of Actionism (Adorno), a barely manageable and complex course is offered, which encourages one to readjust one's attitudes in order to initiate the turnaround the Anthropocene demands. Now, before I myself stray from the path, I break off. Just one more thing, if you are not prepared to get lost, you cannot make maps.

But even events such as these can be and possibly are exclusive and aim to convert the already converted. A more immersive and community-oriented intervention staged by the Theater des Anthropozän / Theatre of the Anthropocene is the *Inventive Expedition Project*, a series of local community events that encourage participants to explore, engage with and experience nature on a micro-level. Initially to be hosted at the tree nursery (Berlin Treptow) and the Späth Arboretum (COVID-19 prevented this format), the project encouraged individuals to document and experience their local treescapes, trees, and sylvan plants through drawing, recording, and feeling the trees in their local vicinity. The mindfulness of the interaction with others human and arboreal species made people more aware of the direct, immediate and local impact of climate destruction. As one participant remarked, "the closer trees get to where humans live, the more hurt and weak they get. With the giant copper birch as an exception" (Theater des Anthropozän 14). Unlike the *Walking Library*, the immediate impact of this intervention was on local policies, on future tree planting, and treescapes.

Lisa Christine Woynarski pushes the idea of eco- and forest-dramaturgy further. Mindful of the anthropocentrism of theatre, performance and the arts in general, her understanding of "ecological anthropomorphism" decentres the human and the human-made world, underscores human embeddedness in ecological systems, and transfers material agency to non-human (or more-than-human) species. To illustrate, Woynarski devised the performance piece *The Celebrated Trees of Nashville, Tennessee* (2012) that attempted to draw attention to the red wiggler worms, a species known for their fertilising capacity. The absence of certain species in the urban landscape, indeed the slow extinction of earthworms globally, becomes the focus of this intervention – a form of public ecological restoration to educate participants/onlookers about the steep decline of earthworm in our soil and the dire consequences this decline has for human and non-human lives.

In her discussion of how to escape the anthropocentrism embedded in theatre and performance, Woynarski references Joseph Beuys' *7000 Eichen – Stadtverwaldung statt Stadtverwaltung* (*7000 Oaks – City Forestation Instead of City Administration*), first presented at the documenta 7 in 1982. The project to plant 7000 oak trees in the city of Kassel, was devised as public (land) art, and at the same time, served to green a city landscape. But, as Beuys pointed out himself, trees have material agency in their environment, they are not only symbols, myth (Forest of Arden), or land art.

> Beuys locates the art and aesthetic experience in biospheric and social terms. By situating the tree as art, he draws attention to the way the tree performs and exercises material agency. To think of Beuys' work as enacting a bioperformativity is necessarily to decentre the human and resist anthropocentrism, recognising the material agency of the more-than-human actants involved (Woynarski, "Ecological Performance Aesthetic" 174).

Richard Powers' monumental tome, *The Overstory* (2018) also taps into this idea of the material agency of trees. (Forgive me for briefly digressing into the world of fiction.) On the one hand, *The Overstory* is a thoughtful reflection on the futility of political/eco-activism – ecological mourning or grief. On the other, the novel is a striking call for humility and a sense of proportion. Indeed, *The Overstory* ends, not with the extinction of mankind but with the reminder that other species, flora and fauna will survive, if possibly changed, and have lived for centuries amongst destruction, war and catastrophe. It is the trees that comment, like a Greek chorus, on the lives of the nine major characters who – with very different backgrounds and backstories – come together as eco-protesters. Reminding us of recent research into the highly successful interconnectedness amongst trees as forms of caring and survival strategy, mankind's existence, even in the history of the universe, seems negligible in the face of the *Sequoiadendron giganteum*.

> The fires will come, despite all efforts, the blight and the windthrow and floods. Then the Earth will become another thing, and people will learn it all over again. The vaults of seed banks will be thrown open. Second growth will rush back in, supple, loud, and testing all possibilities. Webs of forest will swell with species shot through in shadow and dappled by new design. Each streak of colour on the carpeted Earth will rebuild its pollinators. Fish will surge again up all the watersheds, stacking themselves as thick as cordwood through the rivers, thousands per mile. Once the real world ends (233).[4]

When the "real world" ends, another world begins – a post-human world, a non-anthropocentric world that is fertile and abundant. But in his closing words, Powers creates an ambivalent future, a future that is utopian and dystopian simultaneously. "People will learn it all over again" – to destroy, to destruct, to exploit perhaps, to create again a "real" world. Perhaps, Powers seems to suggest, this is the only way we can live in and with the world.

* * *

To conclude and return to the beginning: I began with evoking Frank Kermode, who underscored human nature's need for consonance in concord fictions, to make sense of the beginning, the middle and the end through dramatic, or "daring" *peripeteia*. These forms of narratives are there to find "something out for us, something real," to create meaning. The end times we are in, do not tolerate consonance, and concordance, nor can we evade our anthropocentrism. So where does that leave us with utopia and the performance arts?

---

[4] Sadly, a giant sequoia, commonly (and not appropriately) named "General Sherman" had to be wrapped in aluminium foil in September 2021, to protect it against forest fires.

Luigi Nono's composition, *La lontananza nostalgica utopica futura*, composed in 1988, embodies where utopian performativity is at. Its (former) subtitle "Madrigale a più 'Caminantes' con Gidon Kremer" expresses the peripatetic, unstable, wandering essence of end-times utopianism. The work features a solo violin piece accompanied/disrupted/met by scores and sounds over eight loudspeakers, "irregularly and asymmetrically," "*never near each other*, but in such a way to permit free although *never direct* passage between them" (Nono qtd. in Brodsky 190). The score sets to sound and sensation "the lost sense of direction in utopian imagining" (Mao 19). The possibility of a utopian future embedded in a nostalgic conceptualisation of the past (one that never was), has waned, a utopian future rooted in hope and concordance has been deracinated – we can only witness, wander, search, and practice "infinite listening" as a form of ethical witnessing (Nono qtd. in Griffith). Perhaps then, we can learn it again, the humility and sense of responsibility – humility, after all, is rooted in *humilis*, from earth and soil (see "human-as-humus," Haraway, "Anthropocene" 160).

Works Cited

Angelaki, Vicki. *Theatre & Environment*. London: Macmillan, 2019.
Agamben, Giorgio. "Quando la casa brucia." *Quodlibet*. Web. 5. Oct. 2020. <https://www.quodlibet.it/giorgio-agamben-quando-la-casa-brucia>.
Agamben, Giorgio. "When the House Burns Down." Trans. Kevin Attell. *Diacritics* (2021). Web. 10 Aug. 2021. <https://www.diacriticsjournal.com/when-the-house-burns-down/>.
Abramović, Marina. "What is Performance Art?" *The Artist Is Present Retrospective, MoMA. The Museum of Modern Art Online*. Web. 10. Apr. 2020. <http://moma.org/interactives/exhibitions/2010/marinaabramovic/marina_perf.html>.
Bartleet, Carina E. "Theatre and Performance." *Gender: Nature*. Ed. Iris van der Tuin. Farmington: Macmillan Palgrave, 2016. 297–310.
Bendell, John and Rupert Reed. *Deep Adaptation: Navigating the Realities of Climate Chaos*. London: Polity Press, 2021.
Bloch, Ernst. *The Principle of Hope*. Cambridge, MA: The MIT Press, 1986.
Bregman, Rutger. *Utopia for Realists*. Boston: Little Brown & Co, 2017.
Brodsky, Seth. *From 1989, or European Music and the Modernist Unconscious*. Berkeley: U of California P, 2017.
Colebrook, Claire. "Toxic Feminism: Hope and Hopelessness After Feminism." *Journal of Cultural Research* 14.4 (2010): 323–335.
Coombs, Gretchen. "It's (Red) Hot Outside!" *The Journal of Public Space* 5 (2020): 123–136. Web. 10 August 2021. <https://doi.org/10.32891/jps.v5i4.1407>.
Dolan, Jill. *Utopia in Performance*. Ann Arbor: U of Michigan P, 2005.
Foster, John. Ed. *Facing Up to Climate Reality: Honesty, Disaster and Hope*. London: London Publishing Partnership, 2019.

Fritz, C.E. *Disaster and Mental Health: Therapeutic Principles Drawn from Disaster Studies.* Newark: Disaster Research Centre, U of Delaware, 1996.

Gould, Stephen Jay. *Time's Arrow, Time's Cycle.* Cambridge, MA.: Harvard UP, 1987.

Griffith, Paul. "On Luigi Nono's La lontananza nostalgica utopica futura." *Italian Academy, Columbia University*, 21 Feb. 2018. Web. 10 Apr. 2020. <https://italianacademy.columbia.edu/sites/default/files/devel-generate/phi/Griffiths%20essay.pdf>

Haas, Benjamin. "Hopeless Activism: Performance in the Anthropocene." *Text and Performance Quarterly* 36:4 (2016): 279–296. Web. 10. Apr. 2020. <https://doi.org/10.1080/10462937.2016.1230678>.

Haraway, Donna. "Anthropocene, Capitalocene, Plantationocene, Chthulucene: Making Kin." *Environmental Humanities* 6.1. (2015): 159–165. Web. 10 Apr. 2020. <https://doi.org/10.1215/22011919-3615934>.

Haraway, Donna. *Staying with the Trouble: Making Kin in the Chthulucene.* Durham: Duke UP, 2016. Print.

Harvey, David. *Spaces of Hope.* Berkeley: U of California P, 2000.

Jameson, Fredric. *Archaeologies of the Future: The Desire Called Utopia and Other Science Fictions.* London: Verso, 2007.

Jameson, Fredric. *The Seeds of Time.* New York: Columbia UP, 1994.

Kermode, Frank. *The Sense of an Ending: Studies in the Theory of Fiction.* Oxford: Oxford UP, 1967.

Lehmann, Hans-Thies. *Postdramatisches Theater: Essay.* Frankfurt: Verlag der Autoren, 1999.

Love, Catherine. "From Facts to Feelings: The Development of Katie Mitchell's Ecodramaturgy." *Contemporary Theatre Review* 30.2 (2020): 226–235. Web. 10 Apr. 2020. <https://doi.org/10.1080/10486801.2020>.

Mack, Katie. *The End of Everything: (Astrophysically Speaking).* New York: Scribener, 2021.

Malm, Andreas. *Corona, Climate, Chronic Emergency: War Communism in the Twenty-First Century.* London: Verso, 2020.

Mao, Douglas. *Inventions of Nemesis: Utopia, Indignation and Justice.* Princeton: Princeton UP, 2020.

Marranca, Bonnie. "Robert Wilson and the Idea of the Archive Dramaturgy as an Ecology." *Performing Arts Journal* 15.1 (1993): 66–79. Web. 10 Apr. 2020. <https://doi.org/10.2307/3245799>.

Markussen, Thomas "The Disruptive Aesthetics of Design Activism: Enacting Design between Art and Politics." *Nordic Design Research Conference: Helsinki* (2011). Web. 21 Oct. 2021. <http://www.nordes.org/opj/index.php/n13/article/viewFile/102/86>.

Miéville, China. "Limits of Utopia." *Salvage.* 2015. Web. 10 Oct. 2021. <http://salvage.zone/mieville_all.html>.

Muñoz, José Esteban. *Cruising Utopia: The Then and There of Queer Futurity.* New York: New York UP, 2009.

Out of the Woods. "The Uses of Disaster." *Commune* 5 (22 Oct. 2018). Web. 20 Apr. 2020. <https://communemag.com/the-uses-of-disaster/>.

Out of the Woods. *Hope against Hope: Writings on Ecological Crisis.* New York: Common Notions, 2020.

Patel, Raj, and Jason W. Moore. *A History of the World in Seven Cheap Things.* Oakland: U of California P, 2017.

Powers, Richard. *The Overstory.* London: William Heinemann, 2018.

Raddatz Frank M. "Instruction Manual for a Labyrinth." *Requiem for a Forest.* 2020. Web. 20 Oct. 2021. <https://xn–theater-des-anthropozn-l5b.de/en/requiem-for-a-forest/>.

Robertson, Michael. *The Last Utopians: Four Late Nineteenth-Century Visionaries and Their Legacy.* Princeton: Princeton UP, 2018.

Robinson, Kim Stanley. "Dystopias Now," *Commune* 5 (2 Nov. 2018). Web. 28 Feb. 2022. <https://communemag.com/dystopias-now/>.

Rai, Shirin, et al. "Introduction: Politics and/as Performance, Performance and/as Politics." Ed. Rai Shirin et al. *The Oxford Handbook of Politics and Performance.* Oxford: Oxford UP, 2021. 1–26.

Sanders, Scott R. *Hunting for Hope: A Father's Journeys.* Boston, MA: Beacon Press, 1998.

Sledmere, Maria. "Hypercritique: A Sequence of Dreams for the Anthropocene." *Coils of the Serpent* 8 (2021): 54–79. Web. 20 Apr. 2021. <https://nbn-resolving.org/urn:nbn:de:bsz:15-qucosa2-737006>.

Solnit, Rebecca. *A Paradise Built in Hell: The Extraordinary Communities That Arise in Disaster.* London: The Viking Press, 2009.

Stiegler, Bernard. *The Neganthropocene.* Trans. Daniel Ross. London: Open Humanities Press, 2018.

Theater des Anthropozän. *Dokumentation Erfinderische Expeditionen.* Stiftung AlltagForschung Kunst/Theater des Anthropozän, 2020. Web. 20 Apr. 2021. <https://erfinderischeexpeditionen.alltagforschungkunst.de/wp-content/uploads/Erfinderische_Expeditionen_Doku.pdf>.

Tomšič, Samo. *The Capitalist Unconscious.* London: Verso. 2015.

Walking Library Project. Web. 21 Oct. 2021. <https://walkinglibraryproject.wordpress.com/about/>.

Watt, Ken. "Participatory Performance, Activism, and the Limits of Change." 20 Apr. 2017. Web. 21 Oct. 2021. <https://howlround.com/participatory-performance-activism-and-limits-change>.

Woynarski, Lisa. *Ecodramaturgies: Theatre, Performance and Climate Change.* London: Palgrave Macmillan, 2020.

Woynarski, Lisa Christine. *Towards an Ecological Performance Aesthetic for the Bio-Urban: A Non-Anthropocentric Theory.* PhD. The Royal Central School of Speech and Drama, 2015.

Wright, E.O. *Envisioning Real Utopias.* London: Verso, 2010.

Vicky Angelaki
Environment, Virus, Dystopia: Disruptive Spatial Representations

This essay is concerned with spatial (re)definitions of the dystopian in contemporary British theatre, with emphasis on works that have pursued this thematic thread while, at the same time, reaching for, and, accordingly, developing innovative ways of dramatic representation in the context of what can be described as an unrepresentable, or, at least, not straightforwardly representable crisis context. The primary cause for this is the intersecting domains of climate and social crises – the latter could also be seen to accommodate governmental, or more broadly political, as well as health crises. But there is also, and no less significantly, a crisis that concerns geography and community as well: namely, spatial erosion as part of a systemic, relentless and ethically dubious monetization of place and its offer, whether in terms of soil and landscape, or in terms of aspect, expanse and construction potential. In terms of how we might imagine our surrounding spaces, then, including in their dystopian extensions and iterations, for the purposes of this text I would like to concentrate on two theatre pieces: Martin Crimp's *In the Valley* (2013) and Liz Tomlin's *The Cassandra Commission* (2014).

The two pieces, share, even on the outset, two significant elements: the first concerns their context of creation/production. In the early years of the current century's second decade, even before major political developments that we may now evaluate through considerable hindsight (including the vote for Brexit and the election of Donald J. Trump, as well as the aftermath and backlash of each of these events), experimentation in form, content, and their combinations, driven by artists including the ones on which this essay concentrates, began not only to infiltrate but, also, to shape the norm of contemporary British playwriting. The upset in form that took hold in, especially, the past two decades, and that resolutely moved British playwriting beyond social realism, has proven to be one of the most emphatic and affective ways of capturing and staging – at least to an extent – the manifold trials and tribulations of our time. In the early years of the same decade, or, perhaps, even for the greater part of it, we were still largely referring to the phenomenon that we now primarily identify as "climate crisis" or "climate emergency" as "climate change" – a more innocuous, perhaps, term, that might be seen to ameliorate the event itself; to curb the associated fear; to encourage the possibility of postponement for its truly urgent consideration. Still, in British playwriting there were already voices that were advancing the kind of earnest and pragmatic discourse that was essential so as to deliver a change, shift perspectives, and contribute towards challenging our respective vocabularies.

The period saw the premieres of work I have discussed elsewhere (see *Social and Political Theatre*, *Theatre & Environment*), and will therefore not dwell upon here, and which are important to mention in terms of appreciating the surrounding atmosphere. This included Mike Bartlett's *Earthquakes in London* (2010); Duncan Macmillan's *Lungs* (2011); Stephen Emmott and Katie Mitchell's *Ten Billion* (2012); Duncan Macmillan, Chris Rapley and Katie Mitchell's *2071* (2014), as well as anticipating Caryl Churchill's *Escaped Alone* and Ella Hickson's *Oil* (both 2016). Unlike these pieces (on the understanding that not all of these conform to the following, but that they do combine certain of the characteristics outlined below), however, and this is where the second significant shared characteristic of Crimp and Tomlin's respective texts is to be found, *In the Valley* and *The Cassandra Commission* are neither expansive in their character roster and stage action, or duration, nor structured on scientific data that is, in itself forming the main plot, or, at least, the narratorial pivot of the piece. I would argue that Crimp and Tomlin's texts are driven by a scientific reality, and an event – the climate crisis – which they honour and acknowledge; still, the interpretation and contextualization is left to the audience. In a considerably limited window of time, the pieces, through stage poetry and allusive imagery, construct, and share with the audience, landscapes that we might at a time have conceptualized as fruitful, cared for, inhabited – but which have, in time, suffered exploitation, corrosion, neglect and, ultimately, exhaustion. Such are the landscapes that, as this essay concentrates on showing through engagement with the two short plays, which work disproportionately to their stage duration in terms of the wealth of images that they serve to communicate to the spectator, today we might term "dystopian." In order to reach some appreciation of the dramaturgical complexities on which *In the Valley* and *The Cassandra Commission* hinge, while, at the same time, acknowledging their power of allusion and their expansive visual horizons, I shall concentrate on a theoretical framework shaped by sociological analyses with a focus on the urban landscape, and, specifically, the work of Saskia Sassen. Interrogations of the stage contingents of the climate crisis will be informed by recent discussions that operate, likewise, across disciplinary lines, within the broader fields of drama, theatre and ecocriticism studies.

In their exemplary co-edited volume *Affects in 21st-Century British Theatre: Exploring Feeling on Page and Stage* (2021), Mireia Aragay, Cristina Delgado-García and Martin Middeke curate an excellent array of essays that, in their combined effect, illuminate what has been one of the most rapidly emerging as well as sometimes frustratingly elastic critical and theoretical terminologies. Questions of dramaturgy, structure, experience and impact feature prominently across the volume's essays, and, for the purposes of brevity and clarity, I wish to foreground the work of two colleagues, Mark Robson and Clare Wallace, that is particularly pertinent to the focus of my essay. In his chapter "Theatre at the End of the World," Robson takes on, in analysis that I find equally lucid and poetic, the challenge of notionally approximating and defining the concept of "the future." It is important to note that the future is not singular; that futures proliferate, on the basis of today's actions and inactions, but also on the basis of today's capacities for imaginaries; the future is both here and it is not, both tangible and nebulous. As Robson notes, implicit complexities in our relationship to the future can also foster the need for "[a] species of dramaturgy […] to give shape to events in the world that without the framing power of form may appear simply random or meaningless" (241). Given the focus of this text, not least in concerns of climate crisis representation, imagination is key, or, to quote Robson once more: "To be able to experience the end of the world, it must be possible to project a space or time "beyond" the end, to survive it, at least in imagination – even if that survival may not feel like survival. Such experience is necessarily poetic, made not found, that is, crafted rather than given" (240). I agree fully with this assessment; with the value and agency of the poetic imagination to capturing extreme crisis; with its potential towards representing one of the most slippery and urgent of topics – the environmental emergency.

It is important that without being explicitly written about the climate crisis, Robson's commentary is particularly attuned to it. Here I would specifically like to mention his engagement with dialogues concerning "a complexity around the relation of the future to the end of the world" (240); the "[d]rawing together of affect and catastrophe" (242); and the ways in which affect materializes "in that curious unlocatable location that opens up between performance and audience in a given space" (243). I argue here that, deceptive in their simplicity, their brevity, and, in the case of Crimp's text, also their humour, both *In the Valley* and *The Cassandra Commission* are especially rich in how they manifest such challenges and representational methods. I would, additionally, like to concur with Robson that "[t]o feel the end of the world is to experience that future as hope" (253), a statement that resonates with both texts in terms of content as well as dramaturgy. Both, that is, are written as conversational monologues – they imply an addressee, or multiple addresses, and a compresence; and both begin from a point of understanding that a certain ending of the world as we know it has, already, taken place. It could be one's own world that has collapsed, as in a private catastrophe – or it could be a public, large-scale catastrophe; or it might be, at last, an understanding, that the two are intertwined.

"Well. / Here I am. / I've made it: I've survived. I speak. I move. It's great. It's great to speak," begins Crimp's text (65). There is no accompanying stage direction. Tomlin's work, on the other hand, opens with the stage direction "*A performer steps forward. She holds an A-Z like a witness in a courtroom holds the holy book. She speaks to the audience. She may read from notes*" (183). Then: "It is with some regret that I reflect on my past / It is with some sorrow that I acknowledge my complicity / It is with some degree of guilt that I assume a certain level of responsibility for what has happened and for what will happen next / For what will happen now, here, in this room [...]," we hear (183). The event, then, has both already transpired and is still unfolding. Such is the space, perhaps, for intervention; such is the elasticity of hope that Robson identifies. It is a similar context in Crimp, albeit in a different tone: "And ha ha ha you're thinking – he's telling me what to say. But no. Oh no. Not true. It's me. It's me that speaks. It's me that moves. It's me that thanks you for this opportunity. It's me that says, 'I'll come to the problem later.' Me who survived" (71). To which problem will Crimp's speaker come later, if not the one that has gathered us here, wherever this may be? What might be the indeterminate space that serves as placeholder for the space that used to be? That Tomlin's character holds the map book in the way in which one might hold a Bible, for example, is meaningful: it is landscape that has been hurt; space; the world as we knew it. It is to it that we must at least attempt to make amends; it is its truth that must guide us, even if it no longer appears pertinent – and it is, ultimately, the hubris against nature that has brought about punishment – and repenting, effecting change, may or may not be any longer relevant.

I find the comment of literary scholars David Higgins and Tess Somervell that "Christian tropes – apocalyptic imagery and weather events as divine punishment – demonstrate[...] clear continuities with ostensibly secular discourse around climate change in the twenty-first century" (182) of value to the present discussion. Crimp's text, assuming we read it as a climate crisis narrative, does, after all, make repeated reference to its speaker's encounter with what is, arguably, one of the most representative biblical images: the sheep in the valley. And even though Higgins and Somervell's case study is, in the specific context of the quotation, the work of William Cowper (1785), through an appreciation of how the Romantic imaginary may in fact have been a climate crisis imaginary all along, I find their following assertion highly congruent to the argument of this essay: that an author "looks forward to a future that extends beyond this catastrophic moment [depicted in the work], but is unable or unwilling to describe it in any detail" (185), as they navigate "the stated unfamiliarity of the present" (184), might serve to ground the ensuing discussion.

In the Valley and *The Cassandra Commission*: What Once Was, Can Be No More (?)

"In the midst of this crisis of representation, and in a world that exists 'post-nature' [...], what role can nature writing hope to fulfil?" asks ecocriticism scholar Pippa Marland (290), while also reflecting on how "encounters" (with texts, or contexts, human and other-than-human) "foster an Anthropocene imaginary for the future, enabling us not just to see ourselves as spectres of epochs to come but also, in the shorter term, to imagine alternative human possibilities" (291). I find such enquiries particularly meaningful, especially in the context of how playwriting might work dramaturgically to cultivate affect towards not only heightened audience engagement, but, also, potentially – and, to use the word once more, hopefully – positive change. Such are the through-lines that Clare Wallace has also pursued in her recent work, noting, with a reference to Caryl Churchill's post-2000 theatre, a sustained involvement "with affect and its politics – in the circulation of emotions, the capacity to affect and be affected, the improbable and transversal connections opened by these movements and the warping of cause and effect, power surges and deficits [... as] integral to its aesthetics" (48). The statement, I argue, resonates with the approach that Crimp and Tomlin take in their respective pieces, especially the summation that Wallace offers on Churchill, which I propose is directly applicable to both texts considered here, namely, that "[t]he action of the play is simultaneously slight and seismic" (51).

Crimp and Tomlin's pieces are striking in their poetic power, delivering an affront on the senses, while their compact narrative form does not allow for any kind of stretching. The texts are stark and minimalist, yet open and maximalist in scope. Both Crimp and Tomlin's pieces predate Churchill's *Escaped Alone*, which, since its appearance in January 2016, has rightly dominated discourses in the intertwining concerns of formal experimentation, affect, politics, the climate crisis and broadly recognized social and political British theatre of our time. *Escaped Alone*, as well as its predecessor, *Far Away* (2000), are on a par with Crimp and Tomlin's texts in terms of the strong sensation that while we were focused elsewhere the apocalypse has, in fact, taken place, resolutely and completely, and we are now inhabiting a world that is unrecognizable by any measure that may have served us before, or that has formerly allowed us any solace. These aforementioned plays do not make it their business to provide comfort. On the contrary, they make it their business to expose transgression, at the same time as they also re-enact, with precision yet no pedantry, the path that has reasonably led to the outcome we are now faced with. All the while, nothing has been inevitable, the two pieces appear to suggest. There had been points of possible intervention, inter-awareness, community engagement. It is simply that we chose not to take these paths.

For example, let us consider these characteristic extracts:

> I heard the news. What news? The new news. I clicked on the news. I listened to the newest news. I watched the news. There was the news: I watched it. The news was great. It was great great news. I tried to smile. I thought of my friend. I clicked on my friend. I looked at my friend. I closed my friend. I closed the news. I clicked on my mind: my mind opened. I closed it again. (Crimp 66)

And also:

> You are the first to arrive for the last coach to leave the depot.
> You thought you would be fighting for a seat. Elbowing through the weak and the elderly and those unaccustomed to the brutality of public transport like the old days when you took the train to work. The days when there were trains. The days when there was work. But the depot is as deserted as the city. The city where the light is slowly dying, and everyone is trying to get used to the dark. Because the forest has spread into the city and no-one can see the sky anymore. The street signs have been stolen and no-one can find their way home. A city where everyone is afraid of everyone else and no-one believes in the future. Wherever the people of the city have fled, they have not fled here. Which should perhaps tell you something. Which should perhaps warn you, that you are either too late or too trusting of systems that have long since stopped working. (184)

Having exposed complicity without, at the same time, adopting a tone that is narrowly accusatory, both Crimp and Tomlin's texts reveal the consequences:

> You wait in your seat by the window near the back of the coach for others to arrive. But no-one comes. The coach pulls out into the darkness of the city. You take out your A-Z and follow the road that begins at the tear on page 83. No longer sure which city it refers to, it still comforts you to read it from time to time, trying to match the place names and the angles of the lines on the page with the tracings of your journey. (184–185)

As Tomlin's speaker reflects, so Crimp's persona brings the text to its denouement, reflecting on a mysterious figure (the man with the silver cowboy-hat) that indicts the protagonist with questions that have long since stopped being innocuous, and with concepts whose intensity, as in Tomlin's text, now gathers momentum:

> "Where are the car keys?" he might say – or "What happened to your kitchen table?" "Whose breath exactly" – he'll ask me – "filled the crackling bag?" "Are you quite sure you're alive?" he sometimes says – "and if so, what's making you live?" (72)

Both Crimp and Tomlin's texts thrive in detail: references to staples of a life organized on capitalist terms and punctuated by their material tropes land with cataclysmic frequency to populate the otherwise indeterminate space of either piece. Mentions of places – whether as such, or metonymically, through reference to objects (the map, the car keys, the kitchen table) render them visible in their eerie absence. It is in those moments that the disquiet sets in, as the audience is invited to interrogate what, in fact, has become of these places.

In her sustained engagement with urban environments on an equally local and global scale, while examining how the latter has affected the former, Saskia Sassen is ideally placed to provide a form of answer as to the fate of these spaces. Her discourse serves as applicable both to the present and to the slippery future that is key to the dramaturgy of both pieces engaged with here. In "'A Monster Crawls into the City,'" which occupies that fertile territory between geography and sociology in the context of economy impacts, Sassen offers a discursive hypothesis that is both real and a dystopian fairy-tale at the same time. The colonization of urban space by capitalism, facilitated, at least partially, by locals, becomes revealed as the source of that familiar uniformity that, if we take Crimp and Tomlin's respective texts as our guides, becomes, also, the cause for eventual spatial degradation. As Sassen's text reads:

> In time, the people heard that the same thing was happening in other cities too. They were all beginning to look the same with their tall, identikit buildings. It was as if a monster had crawled into each city and was chomping away at it. Chomp, chomp, chomp.
> "You cannot have a city without neighbourhoods," the people said. "That is where most of us live and shop and go to school." Yet the monster carried on eating neighbourhoods to make room for its tall towers, making a weird kind of tissue for which the people did not quite have a name. The monster called it "urban." This tissue wasn't at all like the tissue of the neighbourhoods or even the old city centre. It was killing the people's houses and small shops and little streets and squares. This was a bad time – there were not enough places to live any more. And as the monster kept coughing out tall towers everywhere, everywhere became nowhere. The people understood that the monster's power was fed by liquid gold. ("A Monster")

It is revealing that, in order to capture a fluid present, whereby space, like the present and future that it might have otherwise served as the physical interconnector for, is slippery, and our orientations, geographies and itineraries as citizens and individuals are at risk, Sassen turns to a genre which could also be described as allegory. The same might be said of Crimp's text, or Tomlin's: we gather, as audiences, to hear stories whereby we begin to find ourselves increasingly implicated, as made possible, in either text, by the device of addressing attendees in a room, making use of the distance of a dramaturgical structure, while challenging it at the same time through the suggestion of civic involvement. That both texts employ a singular persona as a tour guide for the apocalyptic landscape, someone who is both a stand-in for the audience and a privileged observer at the same time – in the sense of presenting knowledge that we may not already possess, is also important in terms of the allegory. Ultimately, we are told a story that, eventually, we realize casts us at its centre, rather than prioritizing the individual speaker we might have initially assumed to be the protagonist. This, too, is realism – but of a novel, disquieting kind. The form and content of Crimp and Tomlin's texts are far from incidental: to capture the radical uncertainties of the present, but, importantly, as that present forever unfolds and new crises invade our everyday, novel textual forms are needed to grasp at the super-enhanced crises of our time.

The "monster" that Sassen talks about, is, of course, neoliberal capitalism – a force that has been as brutal in its transformational powers, as it has been proven fallible; the latter is a point that also emerges later on in Sassen's text. But as with Crimp and Tomlin's, so Sassen's phraseology, in our present moment, is especially prescient and poignant. First, it is capitalism that devours people, cities; it is people and capitalism that devour the environment; then, another, more aggressive force devours everything, and that which we thought was the everyday space – though, important to note, not equally accessible to all; not common; not collective, but transgressed upon – is changed to the unrecognizable. Out is the everyday, in is the dystopian; the dystopian becomes the new "normal." "Normal," itself, is rendered an empty signifier. "And as the monster kept coughing out tall towers everywhere, everywhere became nowhere" – let us note the verb here; observe how "everywhere" becomes "nowhere." Universality comes in the absence of sense; in the loss of coherent and cohesive space. The city becomes hollow. The *A-Z* in *The Cassandra Commission* is pointless, superseded, as are the domestic objects in Crimp's *In the Valley*, no longer utilitarian tools of the everyday, in an anthropocentric scenario that has failed.

Environment and Pandemic

Revisiting Crimp and Tomlin's short plays today, in a context where we have had, violently and unexpectedly, to re-evaluate what the dystopian text and the dystopian space stand to represent, I cannot but conceptualize of the pieces, in addition to their environmental, and, specifically, their climate crisis resonance, as prescient nods, also, to the COVID-19 pandemic. Returning to my early mention of Churchill, both *Far Away* and *Escaped Alone* have operated on such terms, whether we consider them vis-à-vis the natural or the political landscape, or indeed both. The post-apocalyptic aesthetics and locationality of the pieces is, in my view, the very imperative that Crimp and Tomlin's work also puts forward to the spectator. Alongside it, I should like to acknowledge another important imperative that in my view permeates both Crimp and Tomlin's theatre, whether in plays that take a discursive – if monological – form, such as the ones discussed here, or not. This concerns the writers' attention to the dialectical broadly conceived, that is, to creating an elastic space for the audience to enter, and to occupy the structure of the text, as well as the conceptual landscape that it generates. This elasticity has also allowed both texts to travel into our present time and to reflect directly our recent/current predicament: the pandemic again. Its ruined space can, today, also become conceptualized as the empty streets, the abandoned means of public transport, the route to work that no longer was, the city paths that, desolate of the human agent, were being reclaimed by nature, which we saw at the height of the global lockdowns and movement restrictions. In Crimp's text, it is the simple, yet utterly meaningful use of Past Perfect that illuminates the distance between the present and the transgressive past – the one that we are now, however indeterminate this "now" might be, reeling from: "I'd built whole cities in an afternoon, injected the inhabitants and fed them slices of banana" (70), we hear the speaker confess. But what happens when the capitalist chain is broken? The neoliberalist monetization of space, as we have recently established through the pandemic experience, becomes curbed by a phenomenon greater than its instigator (i.e. capitalism) – the virus.

Without in any way discounting, or negating, the grounds of their earlier resonance, since 2020, then, Crimp and Tomlin's texts have, in my reading, become sharply refracted through the lens of the pandemic. Here, too, there is a "before" and "after." It is essential to note, additionally, that the premiere of Crimp's text took place in a medium that came to dominate the staging and/or dissemination of theatrical performance at the time of the eruption and highest impact of COVID-19: digital theatre. In the summer of 2013, as I discuss in previous work ("(Up)Setting the Scene"), and as one of its then newly appointed Artistic Director Vicky Featherstone's first initiatives, the Royal Court Theatre, as part of its *Open Court* summer festival suite of events, featured the series "Surprise Theatre." Under its auspices, theatre events of different formats were presented to a limited audience on site, spectators attending a one-off event they had purchased a ticket for with no prior knowledge as to content. A number of the performances were also broadcast online, and *In the Valley* was one of these.

There is, then, an extraordinary prescience in the piece that materializes in both form and content. For example, let us consider:

> Oh how I loved my God: And oh how the same breath that animated – correction tried to animate – that tried to animate the universe now puffed up then sucked back in the crackling bag!
> This was one in the eye for the sceptics. This was one in the eye for the unpickers – and how I hated them! – of human thread.
> Yes how I detested the unpickers of human thread poking their needles into the human fabric so that the human fabric came apart – who'd stripped human beings of their thread the way that looters in a war zone strip buildings of their copper wire to sell the copper. I'd see them unravel the world.
> I'd watched them – oh yes watched them back in the day winding the world's thread into their ow profitable machines – heard them adopt a tone – listened to them sneer plus listened to them ridicule in best-selling paperbacks the idea of a created world. (Crimp 68–69)

Meanwhile, in Tomlin, at the time after that Past, or that Past Perfect, we encounter:

> The seagulls shriek like the souls of the dead, and circle above the scraps of humanity that litter the beaches rising over the promenade and over the bandstand and up to the entrance to the grand hotel, the road no longer visible under the seaweed and driftwood, the lampposts stunted and their lanterns hanging by a thread. Where the sea starts and the city stops is no longer clear. (187)

In these sections of the text, "[t]*he performer no longer reads from notes. She speaks as if she is describing something that only she can see*" (187). Such a dramaturgical strategy throws spatio-temporal linearity into disarray, in the same vein as Churchill's *Escaped Alone* would accomplish two years later through the shifts in time, space and tone between the segments depicting the four women in the garden and Mrs J's interjecting monologues, which shift us someplace else altogether, however indeterminate. What we may wonder, now, is whether we occupy the space of dystopia, or of prophecy. This might depend on our locationality: that is, whether we consider the pieces retrospectively, or whether we take them in their original production context. But then, it also depends, to return to Tomlin, on whether, in our present moment, we accept to become rooted in the post-climate catastrophe, post-pandemic devastation environment that we inhabit – and intervene where, or if, there may still be scope for intervention. As Tomlin puts this, as her speaker's time and perspective now blend with that of the audience, towards the end of *The Cassandra Commission*: "You wonder if we've blown our second chance. You wonder if our second chance was our last one" (187). We are at the point of the unrecognizable, Tomlin's text informs us. Or, as in Crimp, even God has expired, his breath crackling, and camps no longer matter; what matters is that we observed, when we might have acted, as the capitalist looting of bodies and space was taking place.

But if belief is gone, what thread remains? What thread to connect the human, now that this tissue too, has become devastated, while we hear, as in Tomlin, the souls of the dead shriek? When I argue for "reading back" texts that predate COVID-19 in the context of what has transpired under the pandemic, it is wording like this in intuitive contemporary playwriting that drives the proposition. The sickness of the planet has, finally, caught up to the sickness of the human body. The expiration is analogous, and all-consuming. The texts do not provide answers, yet they compel action through radically affective theatre. Crimp and Tomlin's respective playwriting is distinctly trans-topic and, consequently, invites an intersectional approach towards its understanding. The form captures the intersection: the space (literally and metaphorically) where different crises merge, retaining their individual characteristics while, at the same time, becoming one potent overwhelming scenario, a Pandora's box. I would like, here, to return to Sassen, firstly in her essay "Open Sourcing the Neighbourhood":

> It's important to appreciate the incompleteness of cities – that they can constantly be re-made, for better or for worse, and that they are re-made on their own terms even when the technologies used are similar. Incompleteness and mutability has allowed many of the world's great cities to outlast kingdoms, empires, nationstates, and powerful firms. To take the imagery of incompleteness further: Powerful actors can remake cities in their image. But cities talk back. They do not take it sitting. Sometimes this talking back may take decades, and sometimes it is immediate.

Has an agent of change more immediate than the virus existed in the recent period, one might ask. War, we might answer – but is the virus much different, in its spread, its unremitting claiming of bodies and places? In its rendering of the old familiar enemy – capital – almost quaint and irrelevant? We have talked about globalization before, but its commercialist dystopia now gives way to a viral dystopia – the homogenization remains, but its touchstone is, in COVID times, withdrawal rather than availability.

I would like, similarly, to move to the conclusion of this essay, with a further reference to Sassen. In "The Global City: Introducing a Concept" she notes:

> Global cities around the world are the terrain where a multiplicity of globalization processes assume concrete, localized forms. These localized forms are, in good part, what globalization is about. Recovering place means recovering the multiplicity of presences in this landscape. The large city of today has emerged as a strategic site for a whole range of new types of operations – political, economic, "cultural," subjective. It is one of the nexi where the formation of new claims, by both the powerful and the disadvantaged, materializes and assumes concrete forms. (40)

In the absence of both the powerful and the disadvantaged, which is not to say that COVID is not gendered, or classed, cityscapes have appeared devoid of their operations. "Recovering place means recovering the multiplicity of presences in this landscape," then denying them. The traces, the impacts, are judged by their disappearance. Or, as Crimp's speaker notes, by what, today, we might interpret as radical rewilding, and the dramatic negation of privileged certainty that the virus generated, as sites fell into uncertainty. It is a setting that we might agree could speak to the poetic imagination of the valley that has replaced what was before it:

> But it is a fact that I've seen God. I've seen him in my kitchen. And I've seen him in my entrance hall.
> Back in the day: back in the day, I mean, of kitchens and of entrance halls – before the valley, before the sheep. (67)

Or in Tomlin:

> The coach pulls up as the road runs out. [...] You feel your way down the coach in darkness, the door is already open and as you step out into the gently lapping waves of the sea, the sea that is gathering momentum as it gathers together the detritus of the land [...]. The water is rising in the gutters, spilling out onto the pavements, washing away the cars and signposts as if they are nothing more than plastic toys. (186)

The biblical disaster is no longer a figure of speech in this new landscape, which reclaims itself and, as it recovers, breathing on its own pace, the thick nexus of human absence registers: as one body, collective, and as many, individual. No longer peopled, the place remembers itself. That this startles us, that it frightens us, even, does not mean that we should not have known to expect it.

Conclusion

That *In the Valley* and *The Cassandra Commission* foreground space, as physical, emotional and conceptual territory alike, and that they set this up, both as individual and as collective and – perhaps – shared, with possibilities for the two to interact, is crucial to our understanding of how the pieces are anchored. At the same time, the texts retain an ability to become re-located in each reader/spectator's mind, recast under the light of different crises. Through their elliptical form and expansive field of reference, the short plays navigate (quite literally) unprecedented catastrophe while evidencing that the human endeavour of mapping, as well as of providing linearity and narratives, may have once functioned as a noble incentive, but no longer holds. The romanticism and desirability of a knowable path may linger, as might the artefacts of a past – still relatively recent, enough so to survive in memory quite comfortably, yet no longer making sense – but there is no *A-Z* guidebook to help us through our present dystopias. What remains, is, perhaps, "this capacity to be affected that may flow forward into anger at injustice," as Wallace notes (58). There is also, of course, the possibility for realization that one may be complicit in the injustice, which is where the theatre's interventionist potential is, arguably, at its sharpest.

Works Cited

Angelaki, Vicky. "(Up)Setting the Scene: *Open Court* as Staging and Spectating Intervention." *Contemporary Theatre Review* 25.4 (2015): 472–487.
Angelaki, Vicky. *Social and Political Theatre in 21st-Century Britain: Staging Crisis*. London: Bloomsbury, 2017.
Angelaki, Vicky. *Theatre & Environment*. London: Red Globe Press, 2019.
Aragay, Mireia, Cristina Delgado-García, and Martin Middeke. Eds. *Affects in 21st-Century British Theatre: Exploring Feeling on Page and Stage*. London: Palgrave Macmillan, 2021.
Bartlett, Mike. *Earthquakes in London*. London: Methuen Drama, 2010.
Churchill, Caryl. *Escaped Alone*. London: Nick Hern Books, 2016.
Churchill, Caryl. *Far Away*. London: Nick Hern Books, 2000.
Crimp, Martin. *In the Valley. Writing for Nothing*. London: Faber and Faber, 2019. 61–72.
Emmott, Stephen. *Ten Billion*. London: Penguin, 2013.
Hickson, Ella. *Oil*. London: Nick Hern Books, 2016.
Higgins, David, and Tess Somervell. "Catastrophe." *The Cambridge Companion to Literature and the Anthropocene*. Ed. John Parham. Cambridge: Cambridge UP, 2021. 181–195.
Macmillan, Duncan. *Lungs*. London: Oberon Books, 2011.
Macmillan, Duncan, and Chris Rapley. *2071: The World We'll Leave Our Grandchildren*. London: John Murray, 2015.
Marland, Pippa. "Deep Time Visible." *The Cambridge Companion to Literature and the Anthropocene*. Ed. John Parham. Cambridge: Cambridge UP, 2021. 289–303.
Robson, Mark. "Theatre at the End of the World." *Affects in 21st-Century British Theatre: Exploring Feeling on Page and Stage*. Ed. Mireia Aragay, Cristina Delgado-García, and Martin Middeke. London: Palgrave Macmillan, 2021. 239–256.
Sassen, Saskia. "The Global City: Introducing a Concept." *Brown Journal of World Affairs* 11.2 (2005): 27–43.
Sassen, Saskia. "'A Monster Crawls into the City': An Urban Fairytale by Saskia Sassen." *The Guardian*, 23 Dec. 2015. Web. 13 June 2019. <https://www.theguardian.com/cities/2015/dec/23/monster-city-urban-fairytale-saskia-sassen>.
Sassen, Saskia. "Open Sourcing the Neighbourhood." *Forbes*, 10 Nov. 2013. Web. 13 June 2019. <https://www.forbes.com/sites/techonomy/2013/11/10/open-sourcing-the-neighborhood/?sh=3b0f3d034df0>.

Tomlin, Liz. *The Cassandra Commission. Political Dramaturgies and Theatre Spectatorship: Provocations for Change.* London: Bloomsbury, 2019. 183–188.

Wallace, Clare. "Moving Parts: Emotion, Intention and Ambivalent Attachments." *Affects in 21st-Century British Theatre: Exploring Feeling on Page and Stage.* Ed. Mireia Aragay, Cristina Delgado-García, and Martin Middeke. London: Palgrave Macmillan, 2021. 43–61.

Paola Botham
Towards a Genealogy of the British Feminist Dystopian Play

"I'm just beginning to find out what's possible" is the last line by Marion in Caryl Churchill's first professionally produced stage play, *Owners* (1972). Taken out of context, this statement may sound like a carrier of utopian promise, yet the opposite is the case. After orchestrating a fire where two people die – the man she once loved and his neighbours' child – Marion is not at all sorry. On the contrary, she relishes the fact that she now "might be capable of anything" (Churchill, *Plays: 1* 67). Several decades later a match is lit by Vanessa towards the end of Lucy Kirkwood's first full-length play, *Tinderbox* (2008). After the ensuing explosion kills her lover, Vanessa "*is at the seaside*" and "*smiles*" (98). Both these seminal pieces feature an old-fashioned butcher's shop and a chauvinistic butcher husband, but what brings them together above all else is their anticipation and/or depiction of an ominous future. From a comparison between *Owners* and *Tinderbox*, this chapter aims at establishing the basis for a genealogy of the British feminist dystopian play as a form of political theatre, mapping a parallel territory to that described in the realm of fiction – and science fiction – around the notion of feminist dystopias in women's writing.

Raffaella Baccolini, whose research focuses on utopia and gender studies, identifies the so-called "critical dystopia" as a feminist venture. In her short provocation "Dystopia Matters," from 2006, she suggests that:

> The utopias envisioned by male authors had not been radically different places for women, and through history women had and still have often been citizens of dystopia. The collapse of the western, patriarchal tradition was no big loss for women writers, who at times would even employ irony and detachment to distance themselves from the more regressive and nostalgic views of male writers and to welcome catastrophic scenarios of destruction as a possibility for a clean start. (2–3)

Earlier, in their influential collaboration *Dark Horizons: Science Fiction and the Dystopian Imagination* (2003), Baccolini and Tom Moylan perceived a tendency towards dystopia as typical of twentieth century's literature as a whole.[1] The only interruption in this trajectory corresponds to the "critical utopia" of the

---

[1] There is consensus about this point. Lyman Tower Sargent states it eloquently: "the twentieth century has quite correctly been called the dystopian century, and the twenty-first century does not look much better" ("Do Dystopias Matter?" 10).

1970s, "[s]haped by ecological, feminist and New Left thought" (2). Moylan's definition of the term highlights an "awareness of the limitations of the utopian tradition, so that these texts reject utopia as blueprint while preserving it as dream" (qtd. in Baccolini and Moylan 2). Although the concept of critical utopia "has since been extended beyond this [1970s] time frame" (Sargisson 10), Baccolini and Moylan saw the utopian revival as short-lived, soon to be replaced by a clear "dystopian turn" (3) in the early 1980s, followed by the emergence of more nuanced "critical dystopias" at the end of that decade and thereafter.

Unlike traditional dystopias, which "maintain utopian hope *outside* their pages, if at all," Baccolini and Moylan argue that "critical dystopias allow both readers and protagonists to hope by resisting closure: the ambiguous, open endings of these novels maintain the utopian impulse *within* the work" (7). Concerning form, the authors claim that critical dystopias are driven by "the feminist criticism of universalist assumptions" and, therefore, "resist genre purity in favor of an impure or hybrid text" (7). I will return to the question of whether these parameters can be applied to the context of British theatre, but first it is important to make another theoretical clarification.

In my previous analysis of Caryl Churchill's dystopian drama, centred on her turn-of-the-century play *Far Away* (2000), I insisted on the crucial distinction between dystopia as a progressive endeavour and anti-utopia as a reactionary one (Sotomayor-Botham 170–171). According to Lyman Tower Sargent, while the former is "a non-existent society […] that the author intended a contemporaneous reader to view as considerably worse than the society in which that reader lived," the latter involves "a non-existent society […] that the author intended a contemporaneous reader to view as a criticism of utopianism or of some particular eutopia" ("The Three Faces" 9). Baccolini and Moylan uphold Sargent's differentiation and, in the same volume, Darko Suvin phrases it more bluntly: "The intertext of anti-utopia has historically been anti-socialism, as socialism was the strongest 'currently proposed' eutopia ca. 1915–1975. The intertext of 'simple' dystopia has been and remains more or less radical anti-capitalism" (189).

Gregory Claeys expands on the critique of anti-utopian thinking in *Dystopia(n) Matters* (2013), a collection that – as a welcome contribution to the field – discusses dystopia not only "on the page" but also "on stage." "The demands [utopia] makes respecting the suppression of individuality are justified by the ends achieved in terms of a more just, fair and equal society," Claeys writes. "To its opponents, however, such a view eventuated in Stalinism in all its manifold forms […]. In the most extreme expression of this view, all forms of socialism and social democracy are guilty of these sins" (15–16). Lucy Sargisson, also adopting what she names an "anti-anti-utopian" stand (following Fredric Jameson), makes plain that both "eutopias" and "dystopias" are equal manifestations of utopianism:

> Utopias are radical, in both content and intent. They contain challenges to the roots of contemporary socio-economic and political systems. And they intend to change the world. [...] Within utopias lie "eutopias" and/or "dystopias." These terms help to identify the intention and normative stance of the author or creator of the utopia. Eutopias (from the Greek *eu*: good, pleasing, happy) are intended to be good places: better or perfect societies. They are sometimes referred to as "positive utopias." Dystopias (drawing on the Greek *dys*: bad, wrong, or harsh) are intended to be viewed as bad places (or "negative utopias"). (8)

Like Sargent ("Do Dystopias Matter?" 12) and many others, Sargisson emphasises dystopia's function as a warning, a "jeremiad" (40) or, in the words of Baccolini and Moylan, "the canary in a cage" (2). This interpretation is shared by Dragan Klaić in his pioneering monograph *The Plot of the Future: Utopia and Dystopia in Modern Drama* (1991),[2] where dystopia is depicted as "a gloomy paraphrase of utopia and the last refuge of utopian hope" (7). Siân Adiseshiah's early work on Churchill also attends to the synergy between the "utopian" and the "critical" (or dystopian) impulses: "In transcending the dominant order, that order – vis-à-vis its negation – is implicitly critiqued, and similarly – through critiquing, undermining, and deconstructing the dominant order – the utopian is the silent alternative that necessarily articulates itself" ("Utopian Space" 5).

Even though the 1970s are considered by Baccolini and Moylan as the decade of critical utopian writing in anglophone culture (an assessment mostly based on US fiction), British theatre at the time seems to have already been experiencing a "dystopian turn." Klaić's survey illustrates this trend by including four examples of British stage dystopias from that period: Edward Bond's *Lear* (1971), Tom Stoppard's *Jumpers* (1972), Howard Brenton's *The Churchill Play* (1974, set in the "Orwellian" year 1984 in a prisoner camp named after the former British prime minister) and Howard Barker's *That Good Between Us* (1977). Apart from Stoppard's play, where "[u]topian predictions of the future are shown as mere demagogy" (Klaić 155),[3] these pieces belong to the tradition of political theatre that Merle Tönnies characterises as concentrating "on a perceived power imbalance between the rulers and the subordinate groups in society" and stressing "the ways in which this hierarchy was sustained" (96). Their setting is also "distinctly British" (Tönnies 96), yet it is a vision of Britain where pressing socio-po-

---

2 However, as Trish Reid notes, "there is a striking absence of women" (76) in Klaić's book.
3 Similarly, Stoppard's later trilogy *The Coast of Utopia* (2002) is a critique of the utopian revolutionaries it portrays in nineteenth-century Russia.

litical issues have been extrapolated – or, to use Sargisson's dystopian vocabulary, "stretched to nightmare extremes" (13). Without questioning Tönnies' argument regarding the appearance of "absurdist dystopias" as a 21st-century phenomenon, it is worth noting that the dystopian imagination was already a muscle exercised by British political playwrights in the 1970s.

Looking at the period from 1968 to 1983, John Bull singles out "despair" about political realities as a significant fuel for the "new drama" he describes (10), which Brenton especially translated into memorable theatrical dystopias such as the aforesaid *Churchill Play* and, later, *Thirteenth Night* (1981). However, Brenton also reacted to the difficult decade of the 1980s with "Three Plays for Utopia" at the Royal Court,[4] in another demonstration that "the utopian and the dystopian [can be seen] as two sides of the same coin," as Trish Reid states (77). Although Caryl Churchill is only mentioned in passing by Bull (perhaps because her work was covered extensively in Helene Keyssar's *Feminist Theatre*, a monograph from the same series also published in 1984), her plays epitomise a productive form of political dramaturgy based on the interplay between utopian and dystopian modes. Adiseshiah has explored this subject throughout Churchill's career,[5] yet most scholars have only focused on it since *Far Away*, a play that Eckart Voigts and Merle Tönnies call – with reason – "almost paradigmatic" (297). Nevertheless, Churchill's model of feminist dystopian writing can be traced back at least to her stage debut with *Owners*.

Although *Owners* is not a dystopia in the strict sense of presenting an "alternative near-future [world]" (Reid 76), it does partake in Klaić's definition of the dystopian play as one "in which the future is the explicit or implicit time of action" (2), as further examination will show. Moreover, the (female) baby who dies in the fire anticipates what Elaine Aston and other feminist scholars have noted, that "the child – the girl child in particular – comes to figure throughout Churchill's work as a repository for damaged futures" ("The 'Picasso'" 203). In her aptly named essay "On Churchill and Terror," Elin Diamond makes a similar observation: "Characters, especially children, expressive and often impotent, are aware of distant wars being fought, of bombs that could annihilate them, of anxiety that also threatens [them]" (126–127). One of the most emblematic of these characters is the adolescent Angie in *Top Girls*, who ends the play with

---

[4] This 1988 season featured *Sore Throats* (originally staged in 1979), *Bloody Poetry* (originally staged in 1984) and the premiere of *Greenland*.

[5] The plays she has analysed from this perspective include: *Light Shining in Buckinghamshire* and *Vinegar Tom*, both from 1976 ("Utopian Space"); *Top Girls*, from 1982, and *Fen*, from 1983 ("Utopian Gesture"); *The Skriker*, from 1994, and *Far Away* ("The Dramatisation"); and *Escaped Alone*, from 2016 ("Ageing as Crisis").

the word "frightening" (97). Indeed, for Amelia Howe Kritzer, "*frighten* [is] the most significant single word in Churchill's lexicon" (193).⁶

Churchill, Kirkwood and the Feminist "Family Tree"

Beyond her long-standing involvement in the study of feminist dystopias, Baccolini has also made a useful intervention on the debate about whether women's genealogies are necessary after poststructuralism, defending the link between authorship and gender in the following terms: "Only those who have already been heard [...] can dismiss the notion of voice in a text. For those who have just begun to be heard or whose voices are hardly listened to, a politics of reading for the voice in the text is still indispensable" ("Gender (Still) Matters" 32).

In the specific area of theatre studies, a genealogy of contemporary women's playwriting in Britain was already proposed by Aston more than ten years ago, informed by the increasing importance of affect theory within the field. Her account includes Churchill's play *This is a Chair* (1997), Sarah Kane's *Blasted* (1995) and debbie tucker green's *Stoning Mary* (2005), among others. In the postfeminist climate that enveloped public discourse from the mid-1990s to, possibly, 2012 and the rise of a "fourth wave," Aston maintains that these works are related by an "experiential drive." Thus, they "[lay] claim to a renewal of feminism through the adoption of various dramaturgies and aesthetics that work affectively on audiences so that they might *feel* the loss of feminism, and all [...] that this loss might entail" ("Feeling the Loss" 577).

Despite Churchill's undeniable impact on the landscape of playwriting in general and feminist theatre in particular, this metaphorical family tree is not always visible. Apropos of a quotation about Churchill by her fellow dramatist Moira Buffini, Vicky Angelaki comments:

> I appreciate Buffini's description of Churchill as the playwright "who in the 70s and 80s was the daughter of Brecht, [and now] has become the daughter of Beckett" [...] because it suggests the good intentions of locating Churchill within the respective artistic traditions of two major trailblazers – and certainly there are affinities. However, to think of Churchill as anyone's "daughter" in terms of artistic genealogies risks overlooking the untameable

---

6 Both Diamond and Kritzer locate the start of this dystopian trajectory even before *Owners*. Diamond goes back to Churchill's earliest professional radio play, *The Ants* (1962), whereas Kritzer highlights "the first one-act [piece] produced during her student years at Oxford, *You've No Need to Be Frightened* (1959)" (193).

imagination that has claimed for her the unique position she occupies in contemporary British theatre. (52)

By contrast, Lucy Kirkwood (b. 1984) acknowledges Churchill (b. 1938) as a pivotal figure. When the latter received the Writers' Guild of Great Britain Award for Outstanding Contribution to Writing in January 2018, Kirkwood gave the opening speech. "I keep her works as close as I keep the tea bags and the emergency cigarettes," she said, adding that Churchill's influence is "profound" on "anyone working in the theatre today" ("Cathedrals" 48). Some comparisons have been made between them, especially with reference to Churchill's *Escaped Alone* and Kirkwood's *The Children*, both of which opened at the Royal Court Theatre in 2016. Both these plays put older women centre stage – something that is, unfortunately, still unusual – and both could easily be categorised as dystopian. Commenting on the US transfers of these works, *New York Times* critic Jesse Green described them as offering an "uncanny interweaving of banality and atrocity" (1).[7] With that characterisation in mind, I now return to the earlier pair, *Owners* and *Tinderbox*.

Owners is one of twelve pieces highlighted by veteran critic Michael Billington in his 2020 "Forgotten Plays" newspaper series, focusing on what he deems "lost classics from the last 120 years of theatre" ("Forgotten Plays"). Billington states that, with only one major revival, this is "one of Churchill's least-known works," which "cries out to be rediscovered" ("Forgotten Plays: No 5"). I have argued elsewhere that *Owners* can indeed be considered "as a rich catalogue of Churchill's longstanding preoccupations: capitalist exploitation, gender and sexual politics, identity, ecological damage and the fear of a terrifying future" (Botham 103). That all of this is addressed in the form of a vicious farce may intensify or hamper the play's appeal, depending on taste. For Billington, *Owners* is a "sprightly extravaganza reminding us that Churchill has always been the wittiest of social explorers" ("Forgotten Plays: No 5"). His original review in 1972 was less complimentary, however, criticising the playwright for "throwing everything in [it] bar the kitchen sink" ("Review of *Owners*").

Marion, *Owners*' protagonist, is a property developer with an insatiable appetite for everything: food, clothes, houses, people. As R. Darren Gobert indicates, "she earns her living purchasing Islington properties and selling them off

---

[7] In her recent chapter "Ageing as Crisis on the Twenty-first-century British Stage," Adiseshiah examines the plays in terms of the intersection of these two concepts, concluding that "[w]hile *The Children* maps ageing as personal calamity onto a crisis narrative of environmental apocalypse, *Escaped Alone* [...] regards the two as philosophically distinct" (34). For a different view on Kirkwood's *The Children*, see Farnell.

for profit, thus taking advantage of the area's gentrification (which in the early 1970s was rapid, and which Churchill saw firsthand as a resident)" (46). Marion buys the overcrowded flat where tenants Alec and Lisa live, threatening them with eviction in the process; persuades Alec to rekindle their love affair and then tricks Lisa – who at the start of the play is heavily pregnant with her third child – into signing adoption papers so that the baby boy she has can become the "son" in Marion's husband's business "Clegg and Son." Clegg is a butcher who fantasises about killing Marion because, in his words, he does not accept a wife who "can stand on her own two feet" (8). The play opens in Clegg's butcher's shop when it is about to close – with meat turning grey and smelly – and concludes in "Clegg's new butcher's shop" (6), bought by Marion to keep him appeased.

While this circular spatial arrangement underlines "the inescapable cycle of capitalism" (Botham 105), the time of the piece progresses in a linear and accelerated fashion towards the grim future envisaged in Marion's above-mentioned final words. Therefore, an "implicit" representation of the future – in Klaić terms – gives *Owners* a dystopian character, which is heightened by the increasingly wounding suicide attempts of Worsely, Marion's faithful employee. Also, although there is no suggestion that the action is set at any other point than the present, the fact that "[t]he play takes place in a *developing* bit of North London" (6, emphasis added) implies the fast-moving gentrification process cited by Gobert. Alongside the breakneck pace of Marion, the developer, and Clegg, the butcher, there is another temporal dimension embodied by Alec, who allows for the intended contrast between "western aggressiveness and eastern passivity" (Churchill qtd. in Gooch 40). "Sitting here quietly. Doing nothing" (*Plays: 1* 14) is how Alec chooses to spend his life, until he loses it by becoming "active" in a futile attempt to save a child from the fire. In doing so, he briefly takes the position of "the rebel [who] inevitably fails at the end" of classical dystopias, as identified by Voigts and Tönnies (297) drawing on M. Keith Booker.

The burning of Alec and Lisa's home is intentionally caused by Worsely under separate instructions from Clegg and Marion (who is incensed by her former lover's indifference). The baby girl who perishes with Alec, while Lisa and their older children manage to escape, is the daughter of the middle-class family who has recently bought the property from Marion. After Worsely relates back these events, Clegg tries to appear magnanimous as a way of reasserting authority over his wife: "I forgive you everything," he tells her. "If the police come, say nothing. Leave it all to me. I certainly never intended such a shocking fatal accident as that, and I'm sure you didn't. I will protect you." But Marion's response – quoted here in full – is chilling: "I'm not sorry at all about Alec. Or about that other baby. Not at all. I never knew I could do a thing like that. I might be capable of anything. I'm just beginning to find out what's possible" (67).

Kirkwood's *Tinderbox* is also set in a butcher's shop where the stench of rotten meat is not uncommon. In this case, however, the specific dystopian scenario of a near-future has been activated from the outset, albeit in an ambiguous manner. According to the stage directions, the action occurs "*in Bradford, Yorkshire, sometime in the twenty-first century*" (3), yet British geography has already been altered by climate change and capitalism: Saul (the butcher/husband) and his common-law wife Vanessa have moved North from East London after their area was "flooded so that more salubrious postcodes should be saved" (19). "Hadrian's Channel" (10), stretching thirty-eight miles, now divides England from Scotland and in due course Saul learns, to his nationalistic despair, that Cornwall has been jointly bought by the US and China. Perchik, a young Scottish artist who is on the run from the police after producing a subversive painting of the British prime minister, is recruited by Saul as an assistant and falls for Vanessa when he recognises her as the actress playing Lady Hamilton in an adult film. This feature, *Fellatio Nelson*, was part of "a series of short party-political pornographic films intended to broaden the appeal of the Conservative Party to the masses" (22).

Like Marion, who is reported to have spent time in a psychiatric hospital before starting her business career, Vanessa has experienced mental health problems in the past, yet the tone of grotesque farce is pushed even further here. Saul boasts to Perchik that he has "made" his wife what she is, explaining:

> [D]o you know what "poor Vanessa" did before I began work on her? She soaked her parents' sheets in kerosene and burnt them in their beds. [...] I don't think she realised what she was doing. Vanessa has always been one sausage short of a barbecue. That must be taken into account. But it was down to my testimony that her sentence / was suspended. (57)

Furthermore, the dog-eat-dog theme becomes more than a metaphor in Kirkwood's play. Food is scarce in an England ravaged by riots and looting, so the meat supply in Saul's shop has been complemented for a while by the slaughtering of unsuspecting foreign assistants, before Perchik's arrival. A detective and a delivery man are added to this list of victims during Acts 1 and 2 respectively. Unlike Marion, Vanessa is physically abused by her husband, but the alleged stupidity that reassures him is only a pretence. Once Saul is (almost) accidentally shot by the police and Perchik takes his place, Vanessa's patience finally runs out. When she realises that Perchik – as Saul before – will not fulfil his promise of a move to the coast, she takes matters into her own hands, setting him and the place on fire before reappearing at the seaside, happy to touch the cold water at last.

Given the conspicuous similarities between *Owners* and *Tinderbox* in terms of setting, characters, genre and themes, it is surprising that Churchill was not mentioned by reviewers as a possible inspiration for Kirkwood. Instead, Joe Orton was singled out by many as a model for both plays. In her short preface to *Owners*, Churchill acknowledges that rereading Orton's *Entertaining Mr Sloane* (1963) may have influenced the piece's style. Gobert maintains that this is an understatement and that the text seems "*derivative*" on purpose, deploying an "appropriative method [which] fits a play that argues insistently that 'ownership' is toxic to social relations" (52). Anyhow, the adjective "Ortonesque" was deployed by critics in 1972 to discuss *Owners* and again by Billington when revisiting the play in 2020 ("Forgotten Plays: No 5"). In the much more hostile critical reception encountered by Kirkwood in 2008, at least five reviewers made a comparison with Orton (Billington, "Review of *Tinderbox*"; Spencer; Smith; Koenig; Haydon).

In an exercise almost akin to wine or food tasting, many other references were bandied about in the reviews of *Tinderbox*, not all of them even from the stage. Quentin Letts, for instance, wrote in *The Daily Mail*: "On paper maybe it doesn't sound too bad – Pinter meets Max Frisch by way of Peter Greenaway. In practice it is dreadful: overlong, unfunny, muddled, caricatured" (480). "A slice of *Sweeney Todd* and a dollop of Alfred Jarry, seasoned with *Titus Andronicus*" (480), opined Sam Marlowe in *The Times*. Charles Spencer in *The Daily Telegraph* "spotted [...] Joe Orton, N. F. Simpson, early Pinter, and that modern master of theatrical cruelty Martin McDonagh," while Rhoda Koenig in *The Independent* chose Orton, *Black Mischief*, *Rigoletto* and *Sweeney Todd* (481). "Imagine Joe Orton's *Entertaining Mr Sloane* as rewritten by the League of Gentlemen," invited Andrew Haydon in *Time Out*, yet for *Metro*'s Claire Allfree it was rather "The League of Gentlemen meets *Sweeney Todd*" (482). "*Sweeney Todd* rehashed and embarrassingly immature," judged Kate Bassett in *The Independent on Sunday*, concurring with Susannah Clapp in *The Observer*, who condemned the play thus: "Martin McDonagh without the wit or *Sweeney Todd* without the songs" (481).

It is startling not only that Churchill is absent from this copious inventory, but also that it does not contain any other female authors. This confirms the invisibility of a women's genealogy, even when – whatever one may think of Kirkwood's debut, now superseded by her unquestionable success as a dramatist and screenwriter for more than a decade – *Tinderbox* is a feminist play. Vanessa's moment of realisation makes this evident, when she learns the fate of the real Emma Hamilton after her relationship with Nelson:

I read that book you gave me, Perchik. I read the whole thing. You know what happened to Lady Hamilton? She *died*. In prison. In Calais, Perchik. Her skin went all yellow and she lost her memory and her hair fell out and then she *died*. She never had a powdered wig or nothing. Poor cow. [...] Meanwhile, old Horatio Cyclops gets a state funeral! Well, no thank you. (96)

The Feminist Critical Dystopia on the Stage

Several scholarly contributions on the intersection between feminism and utopianism are instructive for exploring theatrical feminist dystopias as well. The findings of Alessa Johns, who has traced feminist utopian writing from the late Middle Ages until now, can illuminate Churchill's and Kirkwood's focus on childrearing as a social responsibility. Taking the long view, Johns reveals that, in effect, "in feminist utopias this process happens socially, in communities and tribes, rather than strictly within nuclear families" (183). An enduring preoccupation of Churchill, as discussed above, this issue emerges forcefully in *Owners* through Alec's ultimate – and fatal – act of selflessness, which demonstrates his care for other children as much as his own (Botham 108). In *Tinderbox*, Saul and Vanessa's children are already dead at the opening of the play, having been "*slaughtered*" by strangers during a riot (21). While the choice of language links this loss to the very principles of Saul's meat business, the children's names, "Enoch and Maggie" (20),[8] imply their status as "sacrificial" offerings on the altar of their father's ideological war.

Johns also draws attention to environmentalism as a feminist value that predates the modern movement: "Earlier utopian feminists shared this tendency; familiarity with the position of the dependant and the exploited sensitized authors to the plight of abused nature" (189). In this respect, Mary Shelley is notable with "her dystopian science fictions *Frankenstein* (1818) and *The Last Man* (1826), which warn of the consequences of human culture failing adequately to accommodate nonhuman nature" (Johns 190). Through the equation of capitalism, patriarchy and butchery, both plays under consideration – and most of each playwright's subsequent dramatic output – encompass a critique of ecological exploitation. In *Owners*, the target of censure is not only Clegg's industry but also Marion's wholesale embrace of destructive western attitudes: "[T]he animals

---

[8] These are references to the infamous Conservative Member of Parliament Enoch Powell (1912–1998) and the former Conservative prime minister from 1979 to 1990, Margaret Thatcher (1925–2013).

are ours. The vegetables and minerals. For us to consume. We don't shrink from blood. Or guilt. Guilt is essential to progress" (106).

In the realm of theory, and treading in Baccolini and Moylan's footsteps, Ildney Cavalcanti offers a conceptualisation of the feminist critical dystopia in literature that has clear applicability to the stage. Firstly – and refreshingly – she does not make any prescriptions about form, claiming that "feminist literary criticism has evolved past the belief in an inherently feminist aesthetic" (47). Secondly, she appropriates Moylan's term "critical," linking it directly to three issues:

> the negative critique (of patriarchy as well as certain trends in feminist praxis and theory) brought into effect by the dystopian principle; the textual self-awareness in generic terms with regard to a utopian tradition and concerning its own narrative constructions [...] [and] the formation and/or consolidation of a critical-feminist public readership. (48)

Starting from that negative critique, Cavalcanti arrives at a definition of feminist dystopias as "imaginary spaces that most contemporary readers would describe as *bad places for women*, being characterized by the suppression of female desire (brought into effect either by men or by women) and by the institution of gender-inflected oppressive orders" (49, emphasis added). Both *Owners* and *Tinderbox* fit into this definition and, crucially, in Churchill's play Marion is both a victim and an agent of the patriarchal system. Her extradiegetic hospitalisation is narrated by Clegg as an episode in which "they tried to tell her she'd be happier and more sane as a good wife" (10). Yet, in rebelling against that particular oppression, she becomes a fervent representative of the established order, who inflicts its cruel force on Lisa (and Alec). Moreover, the fact that the only alternative to this regime is personified by Alec prevents an essentialist attachment to gender roles. In *Tinderbox*, despite Saul's beliefs, Vanessa is not a hapless spouse either and will eventually exercise her revenge.

"Textual self-awareness" is also an attribute of both pieces, not least in their (conscious?) adoption of an "Ortonesque" tone. In addition, the "hybridity" celebrated by Baccolini and Moylan – what Jane Donawerth calls "genre blending" – makes for a dynamic dramaturgy. In Donawerth's view, "conservative forms are transformed by merging with dystopia, a merge that forces political reconsideration" (29). Satire, employed by both playwrights here, is one of these reactionary styles referenced by Donawerth, inasmuch as "vices and weaknesses are represented in extreme versions to indicate the necessity of change, which [...] is [a] return to the norm or middle ground" (40).

However, as Jonathan Greenberg observes, "[f]eminism and cultural studies have recovered much satiric writing that failed to fit the formulas of Dryden,

Pope, and Swift" (17) and "the transgressive, anti-authoritarian impulses of satire can just as easily make it appear a progressive force that criticizes the powerful and fosters rebellion against the status quo" (23). While the satiric, understood – as Greenberg recommends – as "more nebulous than a genre, [...] an adjective rather than a noun" (10), is seen as distinctive of contemporary British dystopian theatre (Reid 86; Voigts and Tönnies 311), it can also be recognised diachronically in feminist dystopias. According to Cavalcanti, these "paint an exaggerated picture of the existing power relations between the sexes, as if they were placed under a magnifying glass" (53).

Cavalcanti's third factor, "the formation and/or consolidation of a critical-feminist public readership" – or spectatorship, for my purposes – brings back the question posed at the beginning on whether the utopian impulse has to be present *within the work* (as Baccolini and Moylan suggest) in order to manifest itself to the audience of a stage dystopia. Cavalcanti elaborates on this topic through the rhetorical figure of the "catachresis," the general meaning of which is "the use of a word in an incorrect way" (Oxford Dictionary). Such analysis can be applied to make sense of the relationship between dystopia as a form of political theatre and the reality beyond the stage: "dystopias are *overtly* catachrestic because they depict fictional realities that are, to different degrees, discontinuous with the contemporary real (although such realities are drawn in relation to, and as a critique of, the world as we know it)" (Cavalcanti 50).

In the specific case of feminist critical dystopias, Cavalcanti's scheme also engages with the apparent contradiction highlighted by Voigts and Tönnies regarding contemporary dystopian plays, namely, that "a bleak view of the near future as a prevalent mode of political analysis does not necessarily preclude an implicit utopian imaginary" (299–300). Cavalcanti argues that catachresis shows how, in feminist dystopian writing, the utopian object "remains silent, ineffable, in such a way that the paradox of the 'good place' / 'no place' carried by the term *utopia* itself can be maintained" (50). Furthermore, she stresses, "this is crucial in terms of promoting affective identifications and raising readers' political awareness. In other words, the reading position constructed by feminist dystopias denies the satisfaction of desire and reinforces readers' initial position as desiring subjects" (64).

In *Owners*, resistance proves weak against Marion's new reign of absolute dominance. It is she who survives and has the last say, evidencing Churchill's interest in what Mary Luckhurst calls "the politics and operations of complicity" (5; cf. also Adiseshiah, "Utopian Space" 22–23). On the one hand, Churchill's work illustrates that "women are the instruments of other women's and children's oppression: her plays represented from the first what many feminist theorists acknowledged much later – that women are complicit in capitalist structures and can abuse their power just as much as men" (Luckhurst 22). On the other hand, as Luckhurst exemplifies through the ending of *Light Shining in Buckinghamshire* (1976), in a sentence that chimes with Cavalcanti's catachrestic function, "Churchill seems to call herself to silence and reminds her spectators [...] that they are active witnesses, who are being called to question their own complicity with or contestation of politics" (62).

Resistance to the dystopian order is presented more ambivalently in Kirkwood's piece. Perchik, who starts the plot almost in the position of the "visitor" in traditional utopian narratives, becomes entirely complicit towards the end by occupying Saul's patriarchal place.[9] By contrast, Vanessa liberates herself, yet in a violent and individualistic manner. She produces her own happy ending, her own "eutopia" by the sea, while the rest of society – unchanged? – remains hidden from view. However, unlike Marion's horrific act of arson, Vanessa's can elicit sympathy from an audience. This would be the type of affective response signalled by Baccolini at the beginning of this chapter (that is, the permission "to welcome catastrophic scenarios of destruction as a possibility for a clean start") or, as Morgan Lloyd Malcolm puts it in her feminist play *Emilia* (2018),[10]

> That anger that you feel it is yours and you can use it. [...] The house that has been built around you is not made of stone. The stakes we have been tied to will not survive if our flames burn bright. And if they try to burn you, may your fire be stronger than theirs so you can burn the whole fucking house down. (82)

If stage dystopias have a licence to enact our worst nightmares in the present to prevent us from sleepwalking into them in the near future, they can also harbour revolutionary desires. After all, as Johns declares, "gender equality has never fully existed, so it must be imagined" (175).

---

9 Ironically, he also becomes the "hero" (96) in the fictional account Vanessa prepares for the police in order to hide the crime she is about to commit: "I'll say you rescued me from my bed [...] And then [...] you plunged yourself back into the raging fire to attend to the old man who was still snoring while his Empire went up in flames, before finally choking to death on the *fragrant smoke* of his burning flesh" (96–97).
10 My thanks to Ben Poore (University of York) for reminding me of this fiery connection.

Works Cited

Adiseshiah, Siân. "Utopian Space in Caryl Churchill's History Plays: *Light Shining in Buckinghamshire* and *Vinegar Tom*." *Utopian Studies* 13 (2005): 3–26.

Adiseshiah, Siân. "Utopian Gesture in the Cold Climate of Thatcherism: Caryl Churchill's *Top Girls* and *Fen*." *Utopia Matters: Theory, Politics, Literature and the Arts*. Ed. Fátima Vieira and Marinela Freitas. Porto: Editora da Universidade do Porto, 2006. 185–195.

Adiseshiah, Siân. "The Dramatisation of Futureless Worlds: Caryl Churchill's Ecological Dystopias." *Dystopia(n) Matters: On the Page, on Screen, on Stage*. Ed. Fátima Vieira. Newcastle upon Tyne: Cambridge Scholars, 2013. 294–305.

Adiseshiah, Siân. "Ageing as Crisis on the Twenty-first-century British Stage." *Crisis, Representation and Resilience: Perspectives on Contemporary British Theatre*. Ed. Clare Wallace, Clara Escoda, and José Ramón Prado-Pérez. London: Bloomsbury, 2022. 21–37.

Allfree, Claire. "Review of *Tinderbox*." *Metro (London)*, 30 Apr. 2008. Rpt. *Theatre Record* 28.9, 21 Apr.–4 May 2008. 481.

Angelaki, Vicky. *Social and Political Theatre in 21st-Century Britain: Staging Crisis*. London: Bloomsbury, 2017.

Aston, Elaine. "Feeling the Loss of Feminism: Sarah Kane's *Blasted* and an Experiential Genealogy of Contemporary Women's Playwriting." *Theatre Journal* 62.4 (2010): 575–591.

Aston, Elaine. "The 'Picasso' of Modern British Playwrights." *The Theatre of Caryl Churchill*. Ed. R. Darren Gobert. London: Bloomsbury Methuen Drama, 2014. 201–214.

Baccolini, Raffaella. "Dystopia Matters: On the Use of Dystopia and Utopia." *Spaces of Utopia: An Electronic Journal* 3 (2006): 1–4. Web. 20 May 2009. <http://ler.letras.up.pt>

Baccolini, Raffaella. "Gender (Still) Matters: A Genealogy of Feminist Subjectivities." *Women's Voices and Genealogies in Literary Studies in English*. Ed. Lilla Maria Crisafulli, and Gilberta Golinelli. Newcastle upon Tyne: Cambridge Scholars, 2019. 26–40.

Baccolini, Raffaella, and Tom Moylan. "Introduction: Dystopia and Histories." *Dark Horizons: Science Fiction and the Dystopian Imagination*. Ed. Raffaella Baccolini and Tom Moylan. New York and London: Routledge, 2003. 1–12.

Bassett, Kate. "Review of *Tinderbox*." *The Independent on Sunday*, 4 May 2008. Rpt. *Theatre Record* 28.9, 21 Apr.–4 May 2008. 481.

Billington, Michael. "Review of *Owners*." *The Guardian*, 13 Dec. 1972.

Billington, Michael. "Review of *Tinderbox*." *The Guardian*, 29 Apr. 2008. Rpt. *Theatre Record* 28.9, 21 Apr.–4 May 2008. 480.

Billington, Michael. "Forgotten Plays." *The Guardian*. 17 Aug. 2020. Web. 21 June 2021. <https://www.theguardian.com/stage/series/forgotten-plays>.

Billington, Michael. "Forgotten Plays: No 5 – *Owners* (1972) by Caryl Churchill." *The Guardian*. 29 June 2020. Web. 21 June 2021. <https://www.theguardian.com/stage/2020/jun/29/forgotten-plays-no-5-owners-1972-by-caryl-churchill>.

Botham, Paola. "Caryl Churchill." *Modern British Playwriting: The 1970s*. Ed. Chris Megson. London: Methuen, 2012. 99–122.

Bull, John. *New British Political Dramatists: Howard Brenton, David Hare, Trevor Griffiths and David Edgar*. London: Macmillan, 1984.

Cavalcanti, Ildney. "The Writing of Utopia and the Feminist Critical Dystopia: Suzy McKee Charnas's Holdfast Series." *Dark Horizons: Science Fiction and the Dystopian*

Imagination. Ed. Raffaella Baccolini and Tom Moylan. New York and London: Routledge, 2003. 47–67.
Churchill, Caryl. *Plays: 1. Owners, Traps, Vinegar Tom, Light Shining in Buckinghamshire, Cloud Nine*. London: Bloomsbury, 1985.
Churchill, Caryl. *Top Girls*. Ed. Bill Naismith, and Nick Worrall. London: Methuen, 2008.
Churchill, Caryl. "Caryl Churchill." Interviewed by Steve Gooch. *Plays and Players* 20, Jan. 1973. 40.
Claeys, Gregory. "Three Variants on the Concept of Dystopia." *Dystopia(n) Matters: On the Page, on Screen, on Stage*. Ed. Fátima Vieira. Newcastle upon Tyne: Cambridge Scholars, 2013. 14–18.
Clapp, Susannah. "Review of *Tinderbox*." *The Observer*, 4 May 2008. Rpt. *Theatre Record* 28.9, 21 Apr.–4 May 2008. 481.
Diamond, Elin. "On Churchill and Terror." *The Cambridge Companion to Caryl Churchill*. Ed. Elaine Aston, and Elin Diamond. Cambridge: Cambridge UP, 2009. 125–143.
Donawerth, Jane. "Genre Blending and the Critical Dystopia." *Dark Horizons: Science Fiction and the Dystopian Imagination*. Ed. Raffaella Baccolini and Tom Moylan. New York and London: Routledge, 2003. 29–46.
Farnell, Ian. "Utopian Dreams, Dystopian Realities in Lucy Kirkwood and Anne Washburn." *Foundation: The International Review of Science Fiction* 46 (2017): 38–47.
Gobert, R. Darren. *The Theatre of Caryl Churchill*. London: Bloomsbury Methuen Drama, 2014.
Green, Jesse. "Exploring the Half-Life of their Lives." *The New York Times*, 13 Dec. 2017. The Arts/Cultural Desk. 1.
Greenberg, Jonathan. *The Cambridge Introduction to Satire*. Cambridge: Cambridge UP, 2018. 1–52.
Haydon, Andrew. "Review of *Tinderbox*." *Time Out London*, 30 Apr. 2008. Rpt. *Theatre Record* 28.9, 21 Apr.–4 May 2008. 482.
Johns, Alessa. "Feminism and Utopianism." *The Cambridge Companion to Utopian Literature*. Ed. Gregory Claeys. Cambridge: Cambridge UP, 2010. 174–199.
Kirkwood, Lucy. *Tinderbox*. London: Nick Hern Books, 2008.
Kirkwood, Lucy. "Cathedrals Made of Bricks." *American Theatre*, December 2019. 48–49.
Klaić, Dragan. *The Plot of the Future: Utopia and Dystopia in Modern Drama*. Ann Arbor: U of Michigan P, 1991.
Koenig, Rhoda. "Review of *Tinderbox*." *The Independent*, 7 May 2008. Rpt. *Theatre Record* 28.9, 21 Apr.–4 May 2008. 481.
Kritzer, Amelia Howe. *The Plays of Caryl Churchill: Theatre of Empowerment*. London: Macmillan, 1991.
Letts, Quentin. "Review of *Tinderbox*." *The Daily Mail*, 29 Apr. 2008. Rpt. *Theatre Record* 28.9, 21 Apr.–4 May 2008. 480.
Luckhurst, Mary. *Caryl Churchill*. New York and London: Routledge, 2003.
Malcolm, Morgan Lloyd. *Emilia*. London: Oberon Books, 2018.
Marlowe, Sam. "Review of *Tinderbox*." *The Times*, 30 Apr. 2008. Rpt. *Theatre Record* 28.9, 21 Apr.–4 May 2008. 480.
Oxford Dictionary. *Lexico*. Web. 24 Jan. 2022. <lexico.com>
Reid, Trish. "The Dystopian Near-Future in Contemporary British Drama." *Journal of Contemporary Drama in English* 7.1 (2019): 72–88.

Sargent, Lyman Tower. "The Three Faces of Utopianism Revisited." *Utopian Studies* 5.1 (1994): 1–37.

Sargent, Lyman Tower. "Do Dystopias Matter?" *Dystopia(n) Matters: On the Page, on Screen, on Stage*. Ed. Fátima Vieira. Newcastle upon Tyne: Cambridge Scholars, 2013. 10–13.

Sargisson, Lucy. *Fool's Gold? Utopianism in the Twenty-First Century*. Basingstoke: Palgrave Macmillan, 2012.

Smith, Emma. "Review of *Tinderbox*." *The Sunday Times*, 4 May 2008. Rpt. *Theatre Record* 28.9, 21 Apr.–4 May 2008. 481.

Sotomayor-Botham, Paola. "*Lejos*: La distopía socialista-feminista de Caryl Churchill." ["*Far Away*: Caryl Churchill's Socialist-Feminist Dystopia."] *Revista Chilena de Literatura* 83 (2013): 161–175.

Spencer, Charles. "Review of *Tinderbox*." *The Daily Telegraph*, 30 Apr. 2008. Rpt. *Theatre Record* 28.9, 21 Apr.–4 May 2008. 481.

Suvin, Darko. "Theses on Dystopia 2001." *Dark Horizons: Science Fiction and the Dystopian Imagination*. Ed. Raffaella Baccolini and Tom Moylan. New York and London: Routledge, 2003. 187–201.

Tönnies, Merle. "The Immobility of Power in British Political Theatre after 2000: Absurdist Dystopias." *Journal of Contemporary Drama in English* 5.1 (2017): 156–172.

Voigts, Eckart, and Merle Tönnies. "Posthuman Dystopia: Animal Surrealism and Permanent Crisis in Contemporary British Theatre." *Journal of Contemporary Drama in English* 8.2 (2020): 295–312.

Julia Schneider
Performing Utopia? The Contestation of Dystopian Space in Cecelia Ahern's *Flawed* Series

According to Oscar Wilde, "[a] map of the world that does not include Utopia is not worth even glancing at, for it leaves out the one country at which Humanity is always landing. And when Humanity lands there, it looks out, and seeing a better country, sets sail. Progress is the realisation of Utopias" (9). Of course, this paper would never dare to contradict Oscar Wilde, whose genius allowed cucumber sandwiches to become "a metaphor for the superficiality of the late Victorian upper class" (Fenge 93). This paper, however, does dare to argue that utopias can also be realised in performance through the contestation of dystopian space.

Both, utopias and dystopias are by definition spatial concepts. As Gregory Claeys notes, the dystopia designates "a diseased, bad, faulty or unfavorable place" (4), whereas *Oxford Reference* describes utopia as "an imaginary place or state in which everything is perfect." Interestingly, one of the key defining features of dystopia as a literary genre is its tendency to negotiate between utopian and dystopian aspects of societies and hence between utopian and dystopian spaces. Spatial orders in this genre are made to represent social orders, as special spatial realms are reserved for particular classes in the dystopian societies. But while classical dystopias such as *1984* (1949) or *Brave New World* (1932) present bleak futures with relatively stable spatial orders that cannot be disrupted by rebellious tendencies, dystopian fiction for young adults has brought about a shift in the genre of dystopian literature. Its didactic nature encourages ideas about alternative or even "better" ways of living, which are expressed in divergent uses of space by characters who cannot be accommodated properly by the existing spatial order. Being part of contemporary cultural discourses, YA dystopias convey representations which construct specific subject positions and may therefore become sites of identification and inspiration for their readers: "Their narrative techniques often place us close to the action, with first-person narration, engaging dialogue, or even diary entries imparting accessible messages that may have the potential to motivate a generation on the cusp of adulthood" (Basu et al. 1). With their dual purpose to please and instruct, young adult dystopias try to mobilise their readers into action, since, as Tony Coult quite fittingly notes, "[t]he essential characteristic of being a child is the potential to take on board experience, process it, and act upon it – to change" (13).

Hence, in order to show in how far utopia as "the place that is simultaneously a non-place (utopia) and a good place (eutopia)" (Vieira 5) can be created in spatial practice and performance, Cecelia Ahern's YA dystopian *Flawed* series will be investigated. Ahern's dystopian duology is particularly straightforward in its encouragement of political activism in young adults, since its mixed-race protagonist becomes a leader of marginalised citizens in the dystopia. Akin to the #Black Lives Matter movement, which is predominantly led by young Black women (Kendi 503), Celestine's rebellion serves to provide positive images of bodies that are considered "other" – not only in the novels but potentially in the reader's society as well. Her spatial practices disrupt clear-cut boundaries between utopian and dystopian spaces and reveal their heterotopic potential. As counter-sites in which "all the other real sites that can be found within the culture, are simultaneously represented, contested, and inverted" (Foucault, "Other Spaces" 24), heterotopias can question power relations in society – and this is exactly what Celestine does to inspire not only political activism but the creation of utopia.

Cecelia Ahern herself describes her *Flawed* series as a story written "with anger, with love, with passion," a story that demands its readers to see that "none of us are perfect. Let us not pretend that we are. Let us not be afraid that we're not. Let us not label others and pretend we are not the same. Let us all know that to be human is to be flawed, and let us learn from every mistake made so that we don't make them again" (*Flawed* Acknowledgements). Luckily, Ahern also provides us with a definition of what the adjective "perfect" actually means – at least in her dystopia. On the very first page of the second installment, which is itself named *Perfect*, readers are faced with the following explanation: "PERFECT: *ideal, model, faultless, flawless, consummate, quintessential, exemplary, best, ultimate;* (of a person) having all the required or desirable elements, qualities or characteristics; as good as it is possible to be" (*Perfect* Preface). Interestingly, the term "perfect" also frequently appears in approaches to utopia, as the definition by Michel Foucault shows: "They [utopias] present society itself in a perfected form, or else society turned upside down, but in any case these utopias are fundamentally unreal spaces" ("Other Spaces" 24). Yet, since performances always oscillate between imagined spaces and real, social, cultural and political ones, they have the potential to make the unreal space that is utopia graspable and hence "real" for audiences – especially if these performances entail contestations of dystopian space.

Ahern's *Flawed* Series: (De-)Familiarising Racism

The world of Ahern's dystopian duology is marked by a social system which categorises people either as "Perfect" or "Flawed" and, in doing so, mirrors racist practices. The Flawed are "citizens who have made moral or ethical mistakes" (*Flawed* 5) and are therefore ostracised by their communities, which is facilitated by branding parts of these people's bodies with an "F." This signifier visibly settles their status as "the others" and fixes it permanently, so that the majority of society can distance itself from the Flawed and treat them as second-class citizens. Branding the law on bodies is thus taken literally in Ahern's dystopia. However, people who are marked out as Flawed, have not done anything illegal. Their crime consists of carrying out acts that are perceived as damaging to society's norms and values. Still, they are treated worse than "normal" criminals since the latter are free from any charges when they have served their time. A Flawed is Flawed forever – even in death, which is made apparent by the fact that there are special graveyards for Flawed citizens.

Indeed, as is typical of dystopian novels, society in Ahern's dystopia relies on definite and unambiguous classificatory systems and a stable social order. Her society, too, is "[r]igidly hierarchical" and typifies "that civilisation is the new barbarianism" (Albinski 161). These characteristics are also inscribed in and perpetuated through spatial order in the series. Institutions that are supposedly created by and for humanity become hostile. This is why in Ahern's dystopia, Flawed people have extra seats on the bus which are made from red fabric and positioned at the front, facing all the other passengers, "so that everybody on the bus can see that they are Flawed" (*Flawed* 45). Thus, akin to policies of racial segregation, the construction of the bus as a space naturalises their low social status. At the supermarket, too, special Flawed sections reinforce the political and ideological demarcations in society. This seems to suggest that all spaces in the world of the novels are inherently dystopian, since there are always special spatial realms reserved for the Flawed parts of society. As "faulty, defective, imperfect, blemished, damaged, distorted, unsound, weak, deficient, incomplete and invalid" (*Flawed* Preface) people, they are forced to reside in separate spheres which are othered to such an extent that the places are themselves considered Flawed and hence little dystopias.

Yet, the analogy between discriminating against Flawed citizens and racism is not only apparent in the spatial order. Their marginalisation is also based on the fact that their bodies are constructed as contradictory in that they are perceived as exotic and desirable on the one hand, and as threatening and deadly on the other. In the duology, Celestine's classmates invite her to a party only to ambush her upon her arrival. After putting a sackcloth bag over her head and tying her up, they not only kick and beat her, but also strip off her clothes and lock her in a shed to have a look at her brandings: "They picked the one thing that humiliated me most: my body. I never wanted anyone to see it, no one, and yet here I am standing near naked while three people who I thought wanted to be my friends are shining a torch on all the parts of me I can't even bear to look at myself" (*Flawed* 275). Like African women who were displayed publicly as curiosities, Celestine is humiliated, objectified and exposed to the gazes of people who consider her to be "other." The novel thus raises awareness of the atrocious way racist ideologies dehumanise people of colour. The approach of using Flawed people as a group that mirrors the experiences of Black citizens can thereby be understood as a form of defamiliarisation, which is typical of dystopian fiction. According to Maria Varsam, "defamiliarisation makes us see the world anew, not as it is but as it could be; it shows the world in sharp focus in order to bring out conditions that exist already but which, as a result of our dulled perception, we can no longer see" (206). By alluding to racial discrimination without referring to it directly the novels provide new perspectives on the concept of white privilege and encourage readers not only to familiarise themselves with practices of marginalisation but also to question them.

Disrupting Dramaturgical Structures – The Branding Chamber

Celestine North, as "the most Flawed girl in history" (*Flawed* 198), challenges the "official" socio-spatial order by using spaces meant to ostracise Flawed citizens in subversive ways. The Branding Chamber, for example, is meant to spread fear among the public, since it is the very room in which the hot iron brands people as "[s]capegoats for all that is wrong" (94) in society. Built as a small room with an oversized dentist's chair that is facing a floor-to-ceiling pane of glass with two rows of chairs behind it, the Branding Chamber is constructed as a performance space that seeks to lecture through terror. The whole procedure of searing the skin is watched by family members, who are seated behind a window "as though they're at the cinema waiting for the reel to begin" (146). According to Michel Foucault, public executions of penalties not only make people aware of the fact that the slightest offence is likely to be punished, but also arouse feelings of terror by constituting a spectacle of power, in which the spectators take part (*Discipline and Punish* 58). This is also true for Celestine's situation, since she describes the faces of her family members as displaying the terror she feels (*Flawed* 146). Moreover, the separation also prepares her for her new citizen status, since it makes her feel detached from her family: "The excommunication from society is taking effect already within me" (146).

Yet, Celestine contests the space with her anger and bravery because she does not want her family to hear her screams. Instead of crying and feeling sorry for what she did, Celestine decides to show defiance and sticks her tongue out willingly when one of the judges offers her to cancel branding her tongue in exchange for her repentance. Due to her defiance, the judge decides to also brand her spine without anaesthetics – even though it is not part of her original punishment. By afflicting her with physical pain, the judge tries to discipline Celestine's body, but even though she cannot speak since her tongue is swollen from the branding, Celestine stands her ground – despite her inferior power position.

One can hence argue that the Branding Chamber functions as a crisis heterotopia, which is described as a "privileged or sacred or forbidden place[...] reserved for individuals who are, in relation to society [...], in a state of crisis" (Foucault, "Other Spaces" 24). After all, being branded Flawed is a crisis since the whole world shifts for the individual who is branded. This becomes especially apparent in the way the dystopia presents Celestine's first days as a Flawed citizen. Not only are the people who brand her compared to midwives (*Flawed* 147), the chapter describing her first day after the procedure is also introduced with the subtitle "Day One." Celestine's first days as a Flawed are marked by staying in bed and being cared for by her mother: "Day one. I'm home, propped up in my bed by a dozen cushions, organised by Mum, who keeps stepping back to take a look at her work before fluffing and punching again, as if it were a work of art" (159). This equates Celestine with a newborn and thereby illustrates the loss of power effected by the brandings, revealing how powerful the ideologies connected with the brandings are. She ceases to be a subject in other people's eyes, which is also shown in the doctor's visit on her first day as a Flawed citizen. He discloses the results of his examination to her mother but does not address Celestine: "I watch them talk to each other, over me, across my bed, as if I'm not here" (160).

Yet, the chamber also functions as a heterotopia of deviation, which is a space for "individuals whose behavior is deviant in relation to the required mean or norm" (Foucault, "Other Spaces" 25), since only people who do not conform to the values of society are branded. However, Celestine's refusal to adapt to her role in the branding chamber reverses the power structures in the room. It is the judge who gets into a state of crisis and who starts to deviate from society's norms, since his unauthorised decision to brand Celestine's spine is in conflict with the rules. Her performance thus results in the revelation that the head of the Flawed system is Flawed himself, which questions the whole system that has become naturalised for the dystopian society.

The scene in the Branding Chamber furthermore highlights the performativity required for contesting space. The theatricality of the event – the staging, the audience and the ritual of branding somebody Flawed – provides Celestine with the agency to disrupt its dramaturgical structure, since she as the protagonist can reject the role she is expected to play in the plot, i.e. the role of the repentant sinner: "My family will hear me scream. *I* will scream. No. I will not let that happen. I will not allow them to do that to me. [...] It's all part of the fear they place on the public. Let them hear my screams. Make a mistake und you'll end up like her" (*Flawed* 145, 148). Akin to actors, who are judged on the basis of their ability to create a fictional impression of another persona they are meant to embody (White 78), Celestine prepares herself for the brandings and blocks out her fears to create the perfect impression of a strong rebel who is unaffected by the pain she is put through (*Flawed* 156). Because of her audience, she willingly accepts the brandings, which transforms the act of excommunication into a communal event: Celestine's performance leads her audience to identify with her and feel inspired by her: "My family does not sit still. Nor does Mr Berry, who starts thumping on the window, trying to get Crevan's attention. My dad shoves the guard, trying to make him do something to stop this, and they end up having a physical fight in the viewing room. I have never seen my dad like this before" (153). The scene thus underlines the empowering potential of performing space, since it provides Celestine with the agency to undermine apparently fixed power relations in her society.

Reclaiming Home

Another important space in Ahern's dystopia is the "home" – a space which, according to the *OED*, can be defined as the "place where one lives or was brought up, with reference to the feelings of belonging, comfort, etc., associated with it." Dystopias have the tendency to disrupt this idea of a safe space for which people can feel a sense of belonging, since their homes do not offer the characters refuge from their hostile and dysfunctional worlds. This is not only the case in classics such as *1984* in which "home" is equated with constant surveillance, but also in apocalyptic plays such as Edward Bond's *The Children* (2000). In this work a sudden catastrophe has caused a group of children who are friends to set out on a journey in a decaying post-apocalyptic world. Although their situation worsens with every step they take, they do not return home, since their homes cannot offer the emotional or physical support they need. As Claudette Bryanston notes, the play "encapsulates [Bond's] views about young people, and how we are treating them, and not making a 'home' for them in the world" (qtd. in Allen 149).

"Home" in Ahern's dystopia, by contrast, is constructed as an allegedly safe space for Flawed citizens since it is the only place in which they are allowed to conceal their brandings, which gives them a sense of "normalcy": "My home is a cocoon, where the day-to-day of my reality is lived and dealt with, not caring what other people think. I need it to be like this so I can survive, so I can deal with my own reality before hearing other people's twisted perceptions" (*Flawed* 189–190). Yet, the home of Flawed people is constantly invaded by so-called "Whistleblowers," who make sure that the Flawed citizens follow the rules society wants them to live by: "We must eat staple foods, nothing luxurious or fancy, nothing considered unnecessary for our bodies, for our life. Basics. Our intake is measured at the end of every day by a test" (162). In addition to that, rules state that Flawed people have to expect random searches of their private possessions, must adhere to everyday curfews and need to undergo daily lie detector tests. The fact that Celestine has been branded thus transforms her home into a place without privacy; a place in which she is regulated by strangers and feared by members of her own family. Celestine's home can hence be understood as a heterotopia of compensation, which "create[s] a space that is other, another real space, as perfect, as meticulous, as well arranged as ours is messy, ill constructed, and jumbled" (Foucault, "Other Spaces" 27). Akin to colonies, which according to Foucault have "functioned somewhat in this manner," Celestine's house is "absolutely regulated" in order to achieve the "human perfection" ("Other Spaces" 27) she lacks according to society's norms.

Yet, in contrast to the dystopias by Bond and Orwell, Celestine defies the dystopian space her home has become. She convinces the Guild's correspondent Pia Wang of her innocence in the home interviews which were meant to remind her of her "deficiencies" and inspires the journalist to rebel against the Flawed system. Since, as Sargent notes, "[u]topias are generally oppositional, reflecting, at the minimum, frustration with things as they are and the desire for a better life" (8), it can be argued that, by inciting rebellion in an environment that is meant to keep her under control, Celestine succeeds in constructing a utopian space at home that works towards improving life for the marginalised parts of society: "She takes my hands in hers and squeezes tightly; I'm reminded of our first meeting in this room together, the one where we shook left hands so that my branded skin wouldn't touch hers. Now she holds on tightly, my skin against hers. My wound pressed against her smooth skin. It's how it should be, but it moves me deeply" (*Flawed* 308). Moreover, by making a space that is meant to be highly regulated, fully rationalised and purged of all unpredictabilities elude the Guild's control, Celestine transforms this heterotopia of compensation into a heterotopia of illusion, which "exposes every real space, all the sites inside of which human life is partitioned, as still more illusory" (Foucault, "Of Other Spaces" 27).

The act of contesting the dystopian space into which her home has been turned is furthermore highly performative. Celestine refuses to comply with the way the media try to present her and uses them instead to gain more power for her cause by staging herself as a strong young woman who shows no sign of remorse or fear: "She waits for me to break down, to cry, to confess. Instead, I throw my head back and laugh" (*Flawed* 220). Celestine reverses the power positions in the course of the interviews. The information she provides is indeed newsworthy but also dangerous, since the sixth branding she presents to Pia can bring down the whole system in which society believes: "She's trying to act like she doesn't believe me, but I can see her fear" (222). Celestine uses the terror she perceives in the journalist and plays with it by coming closer and thus bringing the very cause of Pia's anxiety – the sixth illegal branding and the truth that lies in it – so close to her that she cannot avoid acknowledging it. To conclude her provocative performance, Celestine even thanks Pia for the "highly informative" interview when she leaves the room (223), as her fear confirms to Celestine how valuable her sixth branding is for her cause; i.e. a more humane society. The demonstration of power with which Celestine closes her performance confronts Pia with the reality that the Guild has constructed for their society and thus makes her doubt the ideologies she has taken for granted. As Richard Schechner notes: "Ending the show and going away also involve ceremony: applause or some formal way to conclude the performance and wipe away the reality of the show, re-establishing in its place the reality of everyday life" (169).

Rebranding Branded Bodies

Yet, it is not only space in terms of physical locations that Celestine contests. The YA dystopia also illustrates how the body as a space can be used for opposition. The F-branding that is placed on the bodies of the Flawed is emblematic of immoral and corrupt behaviour and marks not only their appearance but also their whole character as blemished. As a punishment for aiding a Flawed citizen Celestine receives six different brandings which imprint the laws she was unable to follow onto her body:

> For stealing from society, you will be branded on your right hand. Whenever you go to shake the hands of any decent people in society, they will know of your theft. [...] For your bad judgement, your right temple. [...] For your collusion with the Flawed, for walking alongside them and for stepping away from society, the sole of your right foot. Everytime you connect with the earth, even it will know that you are Flawed to the very root of you. [...] For your disloyalty to the Guild and all of society, your chest, so that if anyone should wish to trust or love you in the future, they will see the mark of your unyielding disloyalty over your heart. And, finally, for the very fact that you lied to this court about your actions, your tongue, so that anyone you speak to or kiss will know that your words fall from a branded tongue and cannot be trusted for the rest of your life. (*Flawed* 141)
>
> Because we have never seen anyone so Flawed to their very backbone like this lady. Brand. Her. Spine. (155)

By constructing six parts of Celestine's body as reminders of her character's corruptness, the Guild as the head of the Flawed system has created little dystopian spaces on her body, which are perceived as "diseased, bad, faulty [and] unfavorable" (Claeys 4) by society. They mark her not only as society's outcast, but also as a threat and warning to the perfect community in which she grew up. This idea is also examined by Michel Foucault, who describes the body as being directly involved in a political field: "power relations have an immediate hold upon it; they invest it, mark it, train it, torture it, for it to carry out tasks, to perform ceremonies, to emit signs" (*Discipline and Punish* 25).

Celestine feels "different to the bone" (*Flawed* 168); however, not because of the physical scars and aches the brandings cause, but because of the "otherness" which has been inscribed onto her body. The letter "F" has been charged with negative associations to such an extent that it has come to represent her whole identity. As a result, she starts to despise and reject her own body:

> I stand naked in front of the mirror, my dressings removed. I hate what I see. My tears fall as my eyes run over the scars on my skin. They have taken away ownership of myself, and they have made me theirs. I want to rip the brandings from my skin. I look away from the mirror. I will never look at myself again. I will never let anyone else see my naked body. Not friends. Not a man. No one. (*Flawed* 181)

At first, Celestine internalises the Guild's conception of herself as a Flawed being and is thus disgusted by her own appearance. This is meant to discipline Flawed citizens into complete obedience. As Foucault notes, by reproducing the crime on the visible body of the criminal the unrestrained presence of the sovereign is confirmed (*Discipline and Punish* 55, 49), which contributes to a functioning system of control. This alludes to the practice of branding slaves, since the scars caused by hot irons "confirmed the presence of the violator, and, therefore, that the body of the enslaved will always be the property of her master and mistress. The enslaved could never claim ownership of her body" (Bakare-Yusuf 317). Akin to the branded body of slaves, Celestine's flesh has become "the signification of her worth within a system whose organizing principle is premised on a proprietary conception of bodies" (Bakare-Yusuf 317).

Yet, Celestine reclaims ownership of her body in the course of the novel. As Elizabeth Grosz claims, the body is not only "a sign to be read, a symptom to be deciphered, but also a force to be reckoned with" (120). Celestine does not accept the inferiority society tries to inscribe on her and actively changes the meanings connected with her scars. By branding the skin above her transversus abdominis as the human body's centre of gravity with a "P" for "Perfect," she turns her body into a powerful tool for attacking the ideologies the government uses to oppress branded people. Being branded both Perfect and Flawed she has constructed herself as balanced since she adds a utopian – a perfect – space to a body that has been constructed as dystopian. She thereby constructs her Black female body as a site for action and protest and thus undermines the idea of it being an object that can be disciplined into inferiority. Here, too, Celestine uses the Branding Chamber as a performance space and lets her ex-boyfriend, the son of the judge who branded her spine, watch the whole act of searing her own body. Creating a scar she did not receive "out of punishment" but "out of pride" (*Perfect* 370), Celestine blurs the clear-cut boundaries between utopian and dystopian spaces. Her body, as a space of manifestation, becomes the mimetic embodiment of the reality of experience – both utopian and dystopian. To some extent she thus builds on the existing perceptions and readings of both her body and the Branding chamber as spaces and then forms a new spatial reality herself. Akin to heterotopias, Celestine has then used her body to represent, contest and invert real sites of her culture (Foucault, "Other Spaces" 24).

Bodies Motioning Change

One of the last acts of contestation in the novel concerns bodies in motion. In order to stop the people who are inspired by Celestine's rebellious acts from rioting, the Guild arranges a parade, in which Flawed people have to walk through the city wearing nothing but a red slip to reveal their flaws to the public:

> Big, small, skinny, fat, black, white, old and young, there is nothing left to the imagination, as we're paraded through narrow cobblestoned streets in front of the audience. [...] This parade was designed to be cruel, to put fear into people's hearts; the public is *supposed* to be horrified. It is a message being sent out to the country: don't believe in the country's ideals and this will happen to you. But nobody can do anything about it – speaking out would be

> to aid a Flawed and they would end up walking alongside us, so everyone keeps their mouths shut, the fear of joining us too great. (*Perfect* 311)

Just as in Caryl Churchill's dystopian play *Far Away* (2000) audiences of Ahern's dystopia are confronted with a monumental scene in which a dystopian space is created by the way in which people are forced to move. Like Churchill's "*ragged, beaten, chained prisoners, each wearing a hat, on their way to execution*" (30), all Flawed citizens are paraded through the streets until they reach Highland Castle as the location where they received their brandings and thus experienced their social death. Essentially, parades, which are commonly defined as "a public march or procession, esp. one celebrating a special day or event" (*OED*), are transformed by Churchill and Ahern to represent violence, shame and disgrace. These scenes stick in their audience's minds as moments of utter inhumanity. The right to move freely and to halt individually is withdrawn in these walks of blame; however, Ahern's dystopia reveals that the parade also entails an automatic connection between the people who are forced to move together. Having reached the courtyard of Highland Castle, all Flawed citizens start to hold hands upon Celestine's request, which emphasises their humanity and inspires other citizens to join their human chain. Past conflicts and ideological demarcations thus give way to a new social arrangement which is not based on the distinction between Flawed and Perfect anymore. Again, a space meant to be dystopian is suffused with utopian elements, so that it becomes the perfectly imperfect location for Celestine's speech of contestation:

> Arrogance, greed, impatience, stubbornness, martyrdom, self-deprecation, self-destruction. These are the seven character flaws Judge Crevan placed on us. But Judge Crevan, there are two sides to every story. When you tell me that I have greed, I call it desire. Desire for a fair and equal society. When you call me arrogant, I call it pride, because my beliefs make me stand above those who oppress me.
> When you say I am impatient, I say that I am daring to question your judgements, which are not law but mere morality courts. You call me stubborn; I say I'm determined. You say I want to make myself a martyr; I say I'm shown *selflessness*. Self-deprecation? Humility. Self-destruction? What I did for Clayton Byrne on the bus was not a deliberate act to ruin my life but a decision based on the belief that what was happening was inhumane. What you see as flaws, Judge Crevan, I see as strengths. (*Perfect* 330–331)

Here again, Celestine's refusal to comply with the prescribed ideologies, and the spatial practices meant to naturalise them, challenges the power structures which have been imposed on her society. The courtyard, which was formerly viewed as the "cobblestones of prejudice" in which people, "who walk over perfect [...] walk back Flawed" (*Flawed* 93), is turned into a heterotopia of illusion "that exposes every real space, all the sites inside of which human life is partitioned, as still more illusory" (Foucault, "Of Other Spaces" 27). All Flawed citizens are indeed reestablished as perfect citizens in this very location; their performance creates utopia.

All in all, Celestine's contestations and reclaimings of space result in a disruption of the spatial order and uncover the utopian qualities and heterotopic cores of the spaces that have been constructed as dystopian. By creating spaces of resistance that celebrate alterity she questions power relations in her society and undermines established spatial classifications. This not only reveals the inequality, subordination and exclusion which has been legitimised on the basis of (apparently) fixed social and spatial structures in her society, but also puts an end to these developments. The injustices that have been imprinted on Celestine's body disempower the Guild's hegemonic system and effect a restructuring of society. With new leaders who abolish the discriminatory laws and construct a political system based on the principles of compassion and logic – notions which Celestine comes to embody in the dystopia – Ahern's duology presents a way out of racist ideologies. As teenage readers are clearly intended to identify with the protagonists of YA dystopias, spatial practices and subversive actions like those presented by Celestine North might inspire them to discover an anti-hegemonial potential in their own spatial practices and "thus literally create space for non-conformist meanings and identities" (Buschmann and Tönnies 9). Thus, even though utopias continue to be formulated in the knowledge that they will never be attained, performances that contest spaces meant to be dystopian can create "effectively enacted utopias" (Foucault, "Of Other Spaces" 24), which can turn utopia into a space which can, after all, be realised.

Works Cited

Ahern, Cecelia. *Flawed*. London: Harper Collins, 2016.
Ahern, Cecelia. *Perfect*. London: Harper Collins, 2017.
Albinski, Nan Bowman. *Women's Utopias in British and American Fiction*. London: Routledge, 1988.
Allen, David. "The Children." *Edward Bond and the Dramatic Child: Edward Bond's Plays for Young People*. Ed. David Davis. Stoke-on-Trent: Trentham, 2005. 149–162.
Bakare-Yusuf, Bibi. "The Economy of Violence: Black Bodies and the Unspeakable Terror." *Feminist Theory and the Body: A Reader*. Ed. Janet Price and Margrit Shildrick. Edinburgh: Edinburgh UP, 1999. 311–323.
Basu, Balaka, et al. "Introduction." *Contemporary Dystopian Fiction for Young Adults: Brave New Teenagers*. Ed. Basu, Balaka; Katherine R. Broad and Carrie Hintz. New York: Routledge, 2013. 1–16.

Buschmann, Heike, and Merle Tönnies. "Introduction: Space In and Beyond Literature." *Spatial Representations of British Identities*. Ed. Heike Buschmann and Merle Tönnies. Heidelberg: Winter, 2012. 7–18.

Churchill, Caryl. *Far Away*. London: Nick Hern Books, 2004.

Claeys, Gregory. *Dystopia: A Natural History: A Study of Modern Despotism, Its Antecedents, and Its Literary Diffractions*. Oxford: Oxford UP, 2017.

Coult, Tony. "Building the Common Future: Edward Bond and the Rhythms of Learning." *Edward Bond and the Dramatic Child: Edward Bond's Plays for Young People*. Ed. David Davis. Stoke-on-Trent: Trentham, 2005. 9–23.

Fenge, Zeynep Z. Atayurt. "Dinner Parties and Power Games in Oscar Wilde's *A Woman of No Importance* and *The Importance of Being Earnest*." *One Day, Oscar Wilde*. Ed. Burçin Erol. Ankara: Hacettepe U, 2016. 81–94.

Foucault, Michel. *Discipline and Punish: The Birth of the Prison*. London: Penguin Books, 1991.

Foucault, Michel. "Of Other Spaces: Utopias and Heterotopias." *Diacritics* 16.1 (1986): 22–27.

Grosz, Elizabeth. *Volatile Bodies: Toward a Corporeal Feminism*. Bloomington: Indiana UP, 1994.

"Home" *Oxford English Dictionary*. Web. 10 Oct. 2021. <https://www.oed.com/view/Entry/87869?rskey=z4lKt0&result=1&isAdvanced=false#eid>.

Kendi, Ibram X. *Stamped from the Beginning: The Definitive History of Racist Ideas in America*. New York: Nation Books, 2016.

"Parade" *Oxford English Dictionary*. Web. 10 Oct. 2021. <https://www.oed.com/view/Entry/137320?rskey=ZDDvXf&result=1&isAdvanced=false#eid>.

Sargent, Lyman Tower. "Utopian Traditions: Space, Time, History." *Utopia: The Search for the Ideal Society in the Western World*. Ed. Gregory Claeys, Lyman Tower Sargent and Roland Schaer. New York: Oxford UP, 2000. 8–17.

Schechner, Richard. *Performance Theory*, rev. ed. New York: Routledge, 1988.

"Utopia" *Oxford Reference*. Web. 7 March 2021. <https://www.oxfordreference.com/view/10.1093/oi/authority.20110803115009560#:~:text=An%20imagined%20place%20or%20state,'%20%2B%20topos%20'place>.

Varsam, Maria. "Concrete Dystopia: Slavery and its Others." *Dark Horizons: Science Fiction and the Dystopian Imagination*. Ed. Raffaella Baccolini and Tom Moylan. New York: Routledge, 2003. 203–224.

Vieira, Fátima. "The Concept of Utopia." *The Cambridge Companion to Utopian Literature*. Ed. Gregory Claeys. Cambridge: Cambridge UP, 2010. 3–27.

White, Graham D. "Compelled to Appear: The Manifestions of Physical Space Before the Tribunal." *Mapping Uncertain Territories: Space and Place in Contemporary Theatre and Drama*. Ed. Thomas Rommel and Mark Schreiber. Trier: Wiss. Verl. Trier, 2006. 73–86.

Wilde, Oscar. *The Soul of Man under Socialism*. 1891. Project Gutenberg, 1997.

Trish Reid
Dystopian Dramaturgies: Living in the Ruins

In the Scottish playwright Morna Pearson's short play *Darklands* (2019), which is set in the north east of Scotland in the year 2045, an infertile couple, Brie and Logan, is offered the opportunity to pioneer a new technology. At the expense of their employer, the corporate and faceless "Company," and upon signing a legally binding agreement, their baby will be "incubated in an artificial womb" (26). They will be able to "watch it grow on a twenty-four-hour live feed" (26). Their daughter, they are assured, will be comprised of "88–93 per cent organic matter" derived from a 50/50 split of their DNA (28). The source of her remaining DNA is never revealed. Brie is more troubled by the offer than Logan, although both welcome the prospect of respite from their current mundane jobs. The disembodied female voice of what we take to be a central computer explains:

> Neither of you will be required to carry on working in your current roles after the gestation period. Your job will be to raise your child as per the agreement. Her progress will be monitored extremely closely, especially at first. As time goes on you'll be expected to make a daily report of any glitches, in case our monitoring methods don't pick them up (30).

Having taken time to consider the more troubling aspects of the proposed arrangement, Brie refuses to sign the agreement but, in a plot twist that highlights her lack of agency, the experiment goes ahead without her consent.

Darklands was commissioned by the National Theatre of Scotland (NTS) and performed in March 2019 – immediately before the first COVID-19 lockdown was imposed in the UK – as part of a trilogy of short plays advertised and published under the umbrella title *Interference*. All three imagine a near-future-world in which digital technologies have transformed work, human interaction and governance. Hannah Kalil's *Metaverse* (2019), the second play in the trilogy, stages an explicitly post-climate catastrophe world in which a female scientist is recruited, again by a powerful company, to work on the development of a virtual reality technology so tactile as to be indistinguishable from real world interaction. The project's progress is charted largely via a series of unsatisfactory virtual interactions between the scientist and her daughter. The trilogy closes with Vlad Butucea's *Glowstick* which is significantly more utopian in its figuring of the interaction of humans with technology, in this instance artificial intelligence (AI). *Glowstick* details the evolving relationship between a severely disabled woman living in a care facility, and IDA, an android assigned her care.

As well as focusing on new technologies, each playwright imagines an outside world in which natural resources and biodiversity have been severely depleted or obliterated, and from which humans have withdrawn into a controlled living environment. By figuring the near future through a series of more or less painful stories set against the very painful backdrop of ecological catastrophe, *Interference* works to reveal dangers threatening in the present and gestures towards potential responses in the future. In this sense it can be usefully thought of as a critical dystopia and can also be seen as part of a larger trend. Representations of environmental disaster or catastrophe are by now ubiquitous within cultural production across a wide range of forms and national boundaries. They are, as Derrick King observes, "no longer limited to mass cultural genres like the post-apocalyptic action film or the disaster sf novel" (195). More particularly, images of ecological apocalypse can now be found with increasing regularity in the theatre, even in that most distinctive strand of UK theatre, new writing. Original plays produced in the last decade or so that deal explicitly with the climate crisis and environmental disaster include Mike Bartlett's *Earthquakes in London* (2010), Emma Adams's, *Ugly* (2010), Moira Buffini, Matt Charman, Penelope Skinner and Jack Thorne's *Greenland* (2011), Dawn King's *Foxfinder* (2011), Richard Bean's *The Heretic* (2011), Lucy Kirkwood's *The Children* (2016), Caryl Churchill's *Escaped Alone* (2016), Stef Smith's *Human Animals* (2016), Ella Hickson's *Oil* (2016) and Clare Duffy's, *Arctic Oil* (2018). This list is by no means exhaustive. In common with *Interference*, a significant number of these plays offer an affective encounter with the near future and, as I have written elsewhere, "taken together they evidence a significant shift in the temporal focus of new writing" which, in the UK at least, has traditionally dealt with topical issues in the present tense (Reid 74). This is not to imply that dystopian tropes and imagery have been absent from new plays set within conventional time frames. Later in this essay I discuss Chris Thorpe's *Victory Condition* (2017) as one example of the dystopian imagination operating in the present tense. Before proceeding with these arguments, I want to set the scene by unpacking the idea of a "politics of ruin" that I refer to in my title, which deliberately references the work of the eminent American political scientist Wendy Brown.

Ruins

In her book *In the Ruins of Neoliberalism: The Rise of Undemocratic Politics in the West* (2019), Brown consolidates her reputation as an acute observer of the contemporary moment by showing how "neoliberal rationality prepared the ground for the mobilisation and legitimacy of ferocious anti-democratic forces in the second decade of the twenty-first century" (7). The anti-democratic forces apparent in the unholy alliance of right-wing authoritarianism and neoliberal economics have been a matter of grave concern for progressives across the globe in recent years, of course, making Brown's latest intervention especially timely. In her earlier book *Undoing the Demos* (2015), which also deals with the threat posed to liberal democracy by neoliberalism, she defined the latter as "an order of normative reason that, when it becomes ascendant, takes shape as a governing rationality extending a specific formulation of economic values, practices, and metrics to every dimension of human life" (30). The 2019 book extends this argument and is built on two key assumptions. First, that neoliberalism needs to be understood as not confined to the economic realm. "Nothing," Brown argues, "is untouched by a neoliberal mode of reason and valuation" and consequently its appraisal requires appreciation of neoliberal political culture and subject "production" (8). As Thomas Biebricher notes in his perceptive review of Brown's book, this means that for Brown "neoliberalism is not understood as an exclusively economizing project but rather as a political one that promotes the duo of markets and morals" (539). It is through this coupling of markets with morals that Brown gets a diagnostic handle on the more reactionary aspects of the contemporary political landscape. In making the connection she builds on the ground-breaking work of Melinda Cooper in *Family Values: Between Neoliberalism and the New Social Conservatism* (2017), which among other things explicitly links "the enormous political activism of American neoliberals in the 1970s" to "the fact of changing family structures" (8). Brown's second assumption – which is most relevant to notions of ruin and ruination – is that while it seems clear that neoliberals and neoliberalism prepared the ground for the ruined political landscape we now inhabit, they are not necessarily its cause, at least not in a straightforward sense. The outcomes of 40-plus years of neoliberal ascendency, she points out, do not only fail to achieve the goals set out in neoliberal textbooks, aspects of them would also be anathema to neoliberalism's founding thinkers. The current alliance of neoliberal economics and the religious right in the US, is one obvious example. For Brown, then, our political landscape is shaped by the *unintended* consequences of the neoliberal project. Our world has suffered a vicious assault in the name of markets, to be sure, but the attack has only been partly successful. A ghastly hybrid that might be described as authoritarian neoliberalism has gained the ascendancy and it is in this sense, according to Brown, that we live in the ruins of neoliberalism. Moreover, as she points out, the liberal/illiberal dichotomy is no longer helpful in this context because figures like Boris Johnson in the UK, and Donald Trump in the US, do not exist in opposition to liberalism but are indeed its very product.

Brown's book deals in detail with neoliberalism as it has developed in the US, and its usefulness for understanding the situation in Europe, and more particularly the UK, is arguably somewhat limited. Indeed, this is one of the arguments Biebricher makes in his widely discussed review of Brown's book. Brown's basic insight, however, that the ruination of democracy is the key aim of contemporary neoliberal politics, holds true and offers a useful focus for this essay, not least because images of the ruin and ruination of democratic values have recurred with noticeable frequency on the contemporary stage, whether in the dystopian near-future play worlds of *Interference*, or in work that deals more directly with the contemporary milieu and its impacts on inter-personal relationships. Particularly useful in this regard is Brown's final chapter "No Future for White Men," which she subtitles "Nihilism, Fatalism, and Ressentiment." In it she draws on Nietzsche and Marcuse to explore the "trivialization and instrumentalization" of values which is "ubiquitous in commercial, political and even religious life today" (*Ruins* 161). For Brown, this process goes beyond the belittling Nietzsche attributes to man's accidental overthrowing of the divine and includes more recent "formations of power that openly defile and defy moral values," and which Nietzsche himself could scarcely have imagined (163). She is worth quoting at length on this topic:

> The paradox of humanly created powers that diminish the human and especially its capacity to shape its world, reaching new intensities just as this capacity is revealed to be all there is [....,] breeds new quantities and subjects of ressentiment [...]. The economizing side of neoliberalism added force to the nihilism of the age and also quickened it, first in leaving nothing untouched by entrepreneurialization and monetization and then, with financialization, submitting every aspect of human existence to investor calculations about its future value. As we become human capital all the way down and all the way in, neoliberalism makes selling one's soul quotidian rather than scandalous (163).

Before returning to the question of how technology is implicated, in the dystopian texts mentioned above, in degradation of democratic values, I want to briefly turn to an example of how the crisis in values is figured in work that focuses on the present day. Chris Thorpe's treatment of these themes in his play *Victory Condition*, which premiered at the Royal Court in 2017, left critics divided and Michael Billington by his own admission "hopelessly baffled" but in its audacious dramaturgy, I want to argue, it encapsulated both the quotidian experience and also the hidden cost of selling one's soul in the ruins of neoliberalism.

Victory Condition

It is Saturday, 21 October 2017. I'm attending a matinee performance of Chris Thorpe's *Victory Condition* with a friend. We have high expectations because Thorpe's previous work has been provocative and productively experimental. In *Confirmation* (2014), for instance, he challenged audience members to examine how confirmation bias impacted their capacity for empathy, especially in their encounter with those who do not share their belief in liberal democracy. This solo performance, which Thorpe made in collaboration with Rachel Chavkin, utilized elements of verbatim, and consisted in Thorpe relating a series of planned encounters with another man who is his "ideological and antagonistic 'enemy,'" a white supremacist neo-Nazi he calls "Glen" (Tomlin 17). Liz Tomlin's perceptive account of the piece in her book *Political Dramaturgies and Theatre Spectatorship: Provocations for Change* (2019), shows that by deliberately choosing a subject outside the "agonistic arena of contestation proposed by Mouffe," *Confirmation* "offers its spectators the opportunity to ask some really difficult questions that are often sidestepped" by theorists of the left including questions about "the role, or ethics, or empathy when dealing with an enemy, rather than an adversary" (161–162). Glen is in many ways a manifestation of the alt-right, his worldview nourished by "neoliberal valorization of libertarian freedom, by wounded, angry white maleness, and by nihilism's radical depression of conscience and social obligation" (Brown, *Ruins* 170). His views are abhorrent, of course, but *Confirmation* is less concerned to paint a nuanced picture of the "enemy" as a means of humanising him, and more interested in demonstrating that the processes of identification and confirmation bias that underpin Glen's strong sense of conviction are mirrored in Thorpe's, and indeed our own. If the experiment is intended to extend the limits of empathic exchange, then, by Thorpe's own admission it is a failure (*Confirmation* 59). For Tomlin, though, this failure of empathy is precisely what opens new ground. The "confrontation with difference, the acknowledgement of the unknowable and the subsequent realization of the fallibility of the subject-self's convictions and identity," she argues, suggests new possibilities for empathic engagement, because it challenges "us to realize that the conviction of being right is an emotional, identitarian commitment," as much as a rational political one (163). Such challenges are rare in the theatre. This explains why my friend and I had high expectations on the afternoon in October 2017, to which I now return.

On the stage of the Jerwood Theatre Downstairs a white box is hanging from scaffolding, which is just visible around the edges. Inside, is a fully furnished open-plan living space. In Chloe Lamford's design it is contemporary, pleasant and a little bland, with "*generic art on the walls*" of the kind chosen "*by a property management company*" (Thorpe, *Victory Condition* 7). A young couple – identified in the published play text simply as woman and man – arrives home from a holiday or perhaps a weekend break. They potter round their flat with easy familiarity, unpacking clothes, sorting out laundry, drinking tea, ordering a pizza, opening a bottle of wine, taking a shower, passing time playing video games. They check their phones and iPads. This physical text is played with clarity and precision by the actors Sharon Duncan Brewster and Jonjo O'Neill. While they engage in these quotidian activities, they talk, but not to each other. Instead, their overlapping monologues are directed at us, the audience. The man begins in a relaxed and friendly register, asserting an affinity with the audience:

> Hello. You have friends. People you care about. People who care about you. So do I [...] You have watched another person dress in the morning while you lay beneath the bedsheets, and they have dressed unselfconsciously in the light slanting through the dirty window, stepping quietly on the wooden boards with their bare feet. They knew you were awake, this other person, but they had no fear of you seeing them naked (9).

Quite quickly, however, although his familiar tone of address does not change, the subject of his monologue turns to the detail of shooting a protestor with a high velocity rifle in a public square: "The maximum time it would take any bullet I fired to hit a target on the surface of the square is, again give or take a tiny amount, nought point two three seconds" (10). The man seems to be a highly trained member of Special Forces. He seems to be working for a repressive regime intent on the assassination of leading protestors. He is soon addressing himself directly to his prospective victim, with whom he identifies strongly and whom he professes to love passionately: "I love you. How I love you. How I wish I could be over there with you. How I wish I could be by your side" (22).

The woman, whose speech is more obviously anxious and stuttering, addresses us as an office worker initially recounting the moment when she arrives at work to find her office frozen in time, or perhaps more accurately buffering. "It looks like the edges have been defined," she tells us, "everything is glowing" (27). Later she thinks maybe she has had a seizure in the underground and never actually reached the office:

> I imagine myself, in the underground station. Through the gaps in the press of people around my body I see a foot twitch. I see piss on my trousers. A newspaper, lying next to me. Fake reports of the end of the world, adverts for unnecessary download speeds. Looming apocalypse and technological overkill. But the end of the world will not be an event. Not in the way we need it to be (29).

The notion that the social fabric is disintegrating, and our world coming to an end, is a recurring theme in woman's monologue, but not in the ways typically figured in disaster movies. "You will not run the streets with your hair on fire," she informs us, nor will you "be narrowly missed by the talon of a monster" (34). "You will not hear the last person with National Authority say God Help Us All" (35). Instead, the signs will be rather more prosaic: "You will substitute ingredients and talk quietly over dinner ... You will laugh when someone says we should have had a revolution ... You will not all die at once ... You will go to work" (35–36).

There is quite a lot going on here. In the first instance, the performance of everyday domestic routine manifest in the physical text is offset by the intercutting of the monologues which signal radical isolation, each character sealed in her/his own world. The fact that the characters speak past and not directly to each other in spite of their obvious intimacy emphasises their alienation. Situated in ill-defined space and time, both monologues are replete with images of contemporary nihilism and fatalism of the kind described by Brown. This is especially clear in Man's reflections on the political regime he serves:

> With the money the President had stolen he built himself a fucking zoo. And we let him. We let him build a fucking zoo and we let him have salt and pepper shakers with solid gold tops and we let him have peacocks by his fucking lake and we let him sign over the rights of business to criminals and we let him invite in organisations that attempted to monetise our debt for the benefit of others and we let him set a terrible example to doctors and police and shop assistants and driving examiners and the guards on the railway and the difference between us is that I see that as one kind of price and you see it as a different kind (22).

Man sees accepting these transgressions as the price of living in a sovereign country where he can buy a drink in a bar, watch television and protect his children, although his victim, he acknowledges, takes "a wider view" (22). His pronounced lack of conscience calls to mind Marcuse's work on "repressive desublimation," on which Brown draws in the final chapter of her 2019 book. Marcuse's thinking is worth summarising here.

In several publications in the 1960s, most famously in *One-Dimensional Man* (1964), Marcuse theorises the advanced state of repressive desublimation in which the subject of commodity culture exists. He defines repressive desublimation as the "flattening out of antagonism between culture and social reality through the obliteration of oppositional, alien and transcendent elements in higher culture by virtue of which it constituted another dimension" (60). Importantly for Marcuse "this liquidation of two-dimensional culture takes place not through the denial and rejection of the 'cultural values,' but through their wholesale incorporation into the established order, through their reproduction and display on a massive scale" (60). As Brown explains, for Marcuse "autonomy declines when comprehension declines […] and comprehension declines when it is not required for survival and when the unemancipated subject is steeped in capitalist commodity pleasures and stimuli" (*Ruins* 167). In his later work *An Essay on Liberation*, Marcuse identified a particularly destructive effect of desublimation in the way "rebellious music, literature, art are easily absorbed and shaped by the market – rendered harmless" (54). The critical response to *Victory Condition*, I would suggest, is enough to show that this is not always the case. Like Billington, the *Evening Standard*'s Fiona Mountford was "left almost totally baffled by" it but for Holly Williams in the *Independent*, Thorpe's "daring use of disconnect between word and deed" dares "you to at least try to make connections." In her *Financial Times* review, Sarah Hemming described the play as an experience that "nags away at you as you walk out of the theatre and into your own bubble." The dramaturgical impulse to dismantle rather than integrate, to expose conflict rather than resolve it, works to unsettle the illusion of unity, which is after all at some level always a political construct. *Victory Condition* reminds us that opacity can be a conscious strategy. It prevents the artwork being consumed too easily.

The clash between routine domestic activity and the extraordinary content of the monologues around which *Victory Condition* is structured, also brings to mind what Louis Althusser – in his homage to Brecht – describes as the tension between the "dialectical temporality" of intelligible historical processes and the "non-dialectical temporality" of the disengaged subject (137–138). As the action unfolds, the gap between rudimentary experience and immediate response on the one hand, and historical explanation on the other deepens and is never resolved. The staging articulates the peculiar combination of disconnect and emotional overload that characterises contemporary experience. In this sense Thorpe offers, I would suggest, a historicizing critique of the paralysed subject in the ruins of neoliberalism.

Victory Condition also offers a ray of hope in its call to "comprehension." This is made particularly clear in a sequence in which the source of the disturbance in Woman's office is revealed to be a young girl who has been trafficked and held captive in another part of the city:

> She lifts her hands from the floor and holds them out in front of her. She moves her hands apart and between them, hanging in the air, a picture of an office […] A woman walks in through a door that fails to swing shut behind her […] And in the space between her hands, which she understands is both smaller and much bigger than the bathroom she has not been able to leave for weeks now, the woman hesitates at the door of the office. And something in the girl makes her want to call out to the woman. To call out, and

say, go on. Find the connection between where you are and I am [...] and do something (40–41).

In spite, then, of the fatalistic tone of much of Man's monologue, the egalitarian sentiment expressed here gestures towards the possibility of a better future. Temporal disjunctions in *Victory Condition* thus work on a number of levels and the fragmentary structure of the performance reflects the ruined history from which it arises.

The title of Thorpe's play draws on the world of gaming, a victory condition being the skills or task one needs to master in order to have definitively won the game. One particularly significant element of video games is their mixing of narrative and game logic, and in recent years number of critics have identified the tension between ludic and narrative structures in video games as ideological spaces. In drawing on this technological metaphor Thorpe acknowledges the centrality of technology to the felt experience of neoliberalism, a theme he shares with the playwrights contributing to *Interference*, although they approach the perceived "problem" of technology more directly.

Technology and Futurity

Some of the critique of technology in *Interference* clearly comes from a place of romantic humanism, which has typically been suspicious of scientific worldviews on the grounds that they reduce man and nature to the status of mere objects. Consequently, so this argument goes, they inhibit the free, independent, and creative individual, in the search for depth of experience and authentic self-expression. It is noticeable in this regard that human agency has all but disappeared in *Darklands* and that human interaction is closely monitored in all three plays by neo-bureaucratic modes of control. The notion that technology has separated, or has the potential to separate, us from some essential part of ourselves, that it alienates us from what makes us human, is thus a shared theme. In these preoccupations the plays chime with contemporary arguments forwarded by thinkers such as Sherry Turkle, who has argued that "technology proposes itself as the architect of our intimacies" and "suggests substitutions that put the real on the run" (1). Similarly, in his study of media advertising *The Attention Merchants* (2016), Tim Wu calls for a "human reclamation project [...] to make our attention our own again, and so reclaim ownership of the very experience of living" (344). We can also perhaps hear echoes of Martin Heidegger, who criticized technology for alienating us, through its disenchanting and instrumentalizing nature, from the mystical experience of being. Technology, in this way of thinking, is seen as impoverishing human life, and as having the dangerous potential to bring ruination to the social fabric. Since I do not believe in the existence of a universal human essence, it is not my intention to defend it against the encroachment of technology in this essay. Instead, I want to highlight aspects of the short plays gathered under the heading *Interference* that point to a more fundamental problem with technology: its role in the reproduction of hierarchies and injustices forced on the majority by neoliberal capitalism.

In an earlier article on "The Dystopian Near-Future in Contemporary British Drama" (2019) my aim was to begin the work of fleshing out a basic taxonomy that might help us to meaningfully distinguish between the types of dystopia utilized in contemporary drama. This endeavour seemed important not least because in searching for a critical lens through which to view this work, I had been struck by the absence of a substantial literature on the topic of theatre and dystopia, or on the question of new writing's relationship with futurity. Siân Adiseshiah's forthcoming *Utopian Drama: In Search of a Genre* (2022) should go some way towards filling this gap. My contention in the 2019 essay was that the dystopian turn in contemporary drama is symptomatic of a particular "structure of feeling," to borrow Raymond Williams' famous formulation, and that it evidences "a profound and dispersed anxiety about the neoliberal present and dissatisfaction with the limitations of realism as a mode for representing it" (Reid 77). In an attempt to make my arguments about the dystopian turn more convincing, I drew specifically on Williams' 1978 essay "Utopia and Science Fiction" because it offers a framework for distinguishing between dystopian narratives. Like most critics, Williams sees the utopian and the dystopian as two sides of the same coin, as "modes of desire or warning in which a crucial emphasis is obtained by the element of discontinuity from ordinary 'realism'" (97). Williams helpfully categorises dystopian narratives into four types, although he concedes that overlaps often occur:

> (a) the hell, in which a more wretched kind of life is described as existing elsewhere;
> (b) the externally altered world, in which a new but less happy kind of life has been brought about by an unlooked-for or uncontrollable natural event;
> (c) the willed transformation, in which a new but less happy kind of life has been brought about by social degeneration, by the emergence or re-emergence of harmful kinds of social order, or of the unforeseen but disastrous of an effort at social improvement;
> (d) the technological transformation, in which the conditions of life have been worsened by technical development. (95)

In my earlier essay I concentrated largely on the "willed transformation" but in what remains of this chapter my intention is to focus on "technological transformation."

Each of the three plays – *Darklands*, *Metaverse* and *Glowstick* – that make up *Interference* deals with applied technology. In each it is technology that has enabled the new life depicted. In this new life human interaction in real time and space is difficult to achieve, and even posited as undesirable – when asked whether she misses touching her husband, one character in Khalil's *Metaverse* responds: "God no. I never liked that stuff. He was always a bit … heavy-handed … Now we can't it's much less … complicated" (Pearson, Khalil, and Butucea 72). In Jen McGinley's design the characters Brie and Logan in Pearson's *Darklands* performed inside Perspex boxes emphasising their physical isolation. Khalil's stage directions are clear in indicating that the characters "*occupy different spaces*" (3). In addition to being characterised by increased isolation, human agency is severely curtailed in these imagined future worlds. In *Darklands* in particular, although Brie and Logan are notionally employees of the "company," they appear to be engaged in what Craig Lambert has called "shadow work," where technology – especially technologically enabled logistics and standardization – has not abolished work but rather proliferated it, particularly in its more mundane, low-paid and repetitive aspects. One of the more shocking revelations in Kate Crawford's *Atlas of AI* (2021), is the presence of vending machines in Amazon's huge fulfilment centre in Robbinsville, New Jersey, which are "stocked with over-the-counter pain killers for anyone who needs them" to combat repetitive strain injury (54). In one wryly satirical sequence in *Darklands*, Logan is "promoted" to a position where he "manages" the circulation of fruit within the facility.

In *Darklands* and *Metaverse* the specific technologies developed – to allow reproduction without intercourse or consent, and to enable virtual reality to replace human interaction – are owned by large companies explicitly seeking profit from their development. "Humans," as Crawford notes, are "the necessary connective tissue" in the development and sustenance of such technologies, but they are not their "most valuable or trusted component" (55). In adopting a dystopian position vis-à-vis technological advance, *Darklands* and *Metaverse* are implicitly critiquing – with more or less acuity and consistency – the accelerationist view that exponential technological development, especially in AI, will surmount the political and social impasses of our current moment, and that we will take a cybernetic leap out of the ruins of late-neoliberalism into what Aaron Bastani has termed *Fully Automated Luxury Communism* (2019). Bastani is not alone in positing this thesis, but he certainly pushes the potential benefits of full automation to their limits, promising a future of boundless leisure for all:

> We will see more of the world than ever before, eat varieties of food we never have heard of, and lead lives equivalent – if we so wish – to those of today's billionaires (189).

The first two plays in *Interference* challenge this utopian reading of technological advance partly by reconnecting it to its capitalist origin and by exposing the relationships of power through which technologies take effect. In this regard, Pearson and Khalil share a number of concerns articulated by Gavin Mueller in his book *Breaking Things at Work: The Luddites Were Right about Why You Hate Your Job* (2021), in which he aims to show how "technology developed by capitalism furthers its goals: compels us to work more" and effectively "limits our autonomy" (7).

The final play in the trilogy, Vlad Butucea's *Glowstick*, engages more directly with prevailing mythologies surrounding artificial intelligence, and unlike Pearson and Khalil embodies technology in the figure of the android. An old woman, River, who is frail, wheelchair-bound and in constant pain, is living in a care facility where she is supported by IDA, "*an AI carebot*" (84). River wishes to find release from her miserable existence in death and we understand she has already attempted to take her own life, which partly explains IDA's presence. The casting of a black actor – Moyo Akandé – in the role of IDA, calls to mind the final section of Caryl Churchill's *Here We Go* (2015), where staging a woman of colour providing personal care to an elderly white patient foregrounds the gendered and racialised character of labour in the existing care sector. At first River is resistant to being looked after by a "machine" and repeatedly asks for a real nurse but we are given to understand, they are on strike. IDA's lack of capacity for empathy is initially foregrounded. She refuses to give River oxygen, insisting the old woman has used up her supply for the hour. River gradually warms to the carebot, largely because she is able to draw IDA into her own world. The play resolves in a flight of fantasy in which, IDA "*takes **River** out through an imaginary door into an imaginary street*" and the "*space slowly turns into a natural environment, full of trees and exotic bird and animal eyes behind every corner*" (102). River's bathtub turns into a lake. IDA experiences the full range of human emotions and however briefly, is "fully alive" (104).

In some ways *Glowstick* exemplifies what Lucy Suchman has described as "the fantasy of the sociable machine" (235). It is a rhetorical gesture that "conjures into existence an imaginative landscape" populated by "socially intelligent artefacts [...] that both think and feel like you and me" (238). Butucea's short play certainly offers a counterbalance to the more technophobic preoccupations of *Darklands* and *Metaverse*. If we understand the posthuman as a merging of biology and technology, it is envisioned in the first two plays as a dystopian condition in which the human/technology merger is posited as a threat to humankind. In *Glowstick* by contrast, Butucea posits what might be thought of as a "utopian enclave" defined by Fredric Jameson in *Archaeologies of the Future*

(2005), as "a space in which new wish images of the social can be elaborated and experimented on" (16).

Conclusion

In a recent article about science fiction and theatre, Ian Farnell notes that "in September 2020, it was announced that, following a successful trial, a humanoid robot named Pepper could soon be rolled out across UK nursing homes to help combat feelings of isolation and loneliness in residents" (373). Farnell reflects on this developing technology and uses *Glowstick* as one of his examples in order to "engage with urgent questions not only around the efficacy of artificial intelligence within caring environments, but with the complex and sensitive matter of care itself" (375). Given the existence of Pepper, the issues explored in *Interference* appear more pressing than we might at first imagine. A growing awareness of the tensions and erasures implicit in contemporary discourses of AI is reflected and inflected, I would suggest, in the dramaturgy of all three plays that comprise *Interference*, although not necessarily in ways that are sustained or rigorous. Both *Darklands* and *Metaverse* hint at the way the vast sums of capital "required to build AI at scale" – as Kate Crawford has shown – mean AI systems "are designed to serve existing dominant interests" (8). Moreover, AI systems cannot exist without "large data sets or predefined rules and rewards" which as we know "depend on a much wider set of political and social structures" (8). Human intelligence cannot be straightforwardly abstracted in the way many of the dominant myths of AI suggest, as if it were something natural and distinct from the social. "In fact," as Crawford reminds us, "the concept of intelligence has done inordinate harm over centuries and been used to justify relations of dominance from slavery to eugenics" (5).

It is worth noting, in conclusion, that *Interference* was performed at City Park, a large office complex in Glasgow's east end which is housed in the former W.D. and H.O. Wills tobacco and cigar factory. This factory employed more than 3,500 people in its heyday and used what was then cutting-edge technology to process tobacco grown mostly in West Virginia. The Wills company was founded in Bristol in 1786, 47 years before slavery was abolished in the British Empire and 79 years before the passage of the Thirteenth Amendment. The building thus casts a shadow, both social and historical, in the landscape of Glasgow's east end. Although it has been repurposed for mixed commercial use, it remains an ambiguous artefact in the contemporary period acting as *memento mori* of past economic and political imperatives that directed both its construction and its decay.

Works Cited

Adiseshiah, Siân. *Utopian Drama: In Search of a Genre*. London: Bloomsbury, 2022.
Althusser, Louis. "The 'Piccolo Teatro': Bertolazzi and Brecht. Notes on a Materialist Theatre." Trans. Ben Brewster. *For Marx*. London: Verso, 1990, 129–151.
Bastani, Aaron. (2019) *Fully Automated Luxury Communism*. London: Verso, 2019.
Biebricher, Thomas. "Book Review: Wendy Brown, In the Ruins of Neoliberalism: The Rise of Antidemocratic Politics in the West." *Perspectives on Politics* 18.2 (2020): 539–541.
Billington, Michael. "*Victory Condition* Review." *Guardian*, 9 Oct. 2017. Web. 14 Nov. 2021 <https://www.theguardian.com/stage/2017/oct/09/victory-condition-review-royal-court>.
Brown, Wendy. *Undoing the Demos: Neoliberalism's Stealth Revolution*. New York: Zone Books, 2015.
Brown, Wendy. *In the Ruins of Neoliberalism: The Rise of Antidemocratic Politics in the West*. New York: Columbia UP, 2019.
Cooper, Melinda. *Family Values: Between Neoliberalism and the New Social Conservatism*. New York: Zone Books, 2017.
Crawford, Kate. *Atlas of AI: Power, Politics, and the Planetary Costs of Artificial Intelligence*. New Haven: Yale UP, 2021.
Farnell, Ian. "Theatre, Science Fiction, and Care Robots: Embodying Contemporary Experiences of Care." *Theatre Journal* 73.3 (2021): 373–389.
Hemming, Sarah. "*Victory Condition*: An Enigmatic Drama at the Royal Court, London." *Financial Times*, 10 Oct. 2017. Web. 14 Nov. 2021. <https://www.ft.com/content/38e416d6-acd8-11e7-aab9-abaa44b1e130>.
Jameson, Fredric. *Archaeologies of the Future: The Desire Called Utopia and Other Science Fictions*. London: Verso, 2005.
King, Derrick. "From Ecological Crisis to Utopian Hope: Kim Stanley Robinson's Science in the Capital Trilogy as Realist Critical Dystopia." *Extrapolation* 56.2 (2015): 195–214.
Lambert, Craig. *Shadow Work: The Unseen, Unpaid Jobs That Fill Your Day*. Berkeley: Counterpoint, 2015.
Marcuse, Herbert. *One-Dimensional Man*. Boston: Beacon Press, 1964.
Marcuse, Herbert. *An Essay on Liberation*. London: Penguin, 1972.
Mueller, Gavin. *Breaking Things at Work: The Luddites Were Right About Why You Hate Your Job*. London: Verso, 2021.
Mountford, Fiona. "*Victory Condition*, Theatre Review." *Evening Standard*. 10 Oct. 2017. Web. 14 Nov. 2021. <https://www.standard.co.uk/culture/theatre/victory-condition-theatre-review-ponderous-show-isnt-a-winner-a3683291.html>.
Pearson, Morna, Hanna Khalil, and Vlad Butucea. *Interference* [*Darklands, Metaverse, Glowstick*] London: Methuen, 2019.
Reid, Trish. "The Dystopian Near-Future in Contemporary British Drama." *Journal of Contemporary Drama in English* 7.1 (2019): 72–88.
Suchman, Lucy. *Human-Machine Reconfigurations*. Cambridge: Cambridge UP, 2007.
Tomlin, Liz. *Political Dramaturgies and Theatre Spectatorship: Provocations for Change*. London: Bloomsbury, 2019.
Thorpe, Chris. *Confirmation*. London: Oberon Books, 2014.
Thorpe, Chris. *Victory Condition*. London: Oberon Books, 2017.

Turkle, Sherry. *Alone Together: Why We Expect More from Technology and Less from Each Other*. New York: Basic Books, 2013.
Williams, Holly. "*Victory Condition*, Royal Court, London, Review." *Independent*, 11 Oct. 2017. Web. 14 Nov. 2021. <https://www.independent.co.uk/arts-entertainment/theatre-dance/reviews/victory-condition-royal-court-review-a7995206.html>.
Williams, Raymond. "Utopia and Science Fiction." [1978] *Tenses of the Imagination: Raymond Williams on Science Fiction, Utopia and Dystopia*. Ed. Andrew Milner. Bern: Peter Lang, 2010. 93–112.
Wu, Tim. The Attention *Merchants*. London: Atlantic Books, 2017.

Luciana Tamas
A Description of This World as if It Were a Beautiful Place: From Avant-Garde Destruction to Dys(u)topias

1 The Negative as an Aesthetic Category

"If," William James begins one of his lectures, "we were to ask the question: 'What is human life's chief concern?,' one of the answers we should receive would be: 'It is happiness'" ("Religion" 78). In this and his following lecture, "The Sick Soul," James draws a parallel between those whom he terms – as the titles anticipate – the "healthy-minded" and the "sick-minded," who deal with the core existential questions in diametrically different ways. Thus, in the "healthy-minded"

> happiness is congenital and irreclaimable. "Cosmic emotion" inevitably takes in them the form of enthusiasm and freedom. [...] I mean those who, when unhappiness is offered or proposed to them, positively refuse to feel it, as if it were something mean and wrong. We find such persons in every age, passionately flinging themselves upon their sense of the goodness of life, in spite of the hardships of their own condition, and in spite of the sinister theologies into which they may be born. ("Religion" 79)

While the "healthy-minded" curate their vision so as to "deliberately [exclude] evil from its field" (88), the "sick-minded" cannot "so swiftly throw off the burden of the consciousness of evil, but are congenitally fated to suffer from its presence" ("Sick Soul" 133–134). There are, however, various degrees of "healthy-mindedness" or "sickness": "Just as we saw that in healthy-mindedness there are shallower and profounder levels [...] so also are there different levels of the morbid mind" (134). Through this psychological partition, James tries not solely to describe two opposing strains in human temperament, but also to define humans' attitude towards the religious or the mystical. In this, notably, he also discusses a series of writers who leaned, in his view, towards one or the other of the two sides of the spectrum. Examples he prefers in the class of the "healthy-minded" are Emerson or Whitman ("Religion" 81, 84–87), while among the "sick-minded" he mentions Tolstoy or Luther ("Sick Soul" 149–157, 137–138). Of the "healthy-minded" he notes that "[o]ne can but recognize in such writers as these the presence of a temperament organically weighted on the side of cheer and fatally forbidden to linger, as those of opposite temperament linger, over the darker aspects of the universe" – and he concludes that their "capacity for even a transient sadness or a momentary humility seems cut off from them as by a kind of congenital anaesthesia" ("Religion" 83). Yet while the "healthy-minded," through their very nature, invariably lead the more fulfilling and "happy" lives, it is, James notes, the "sick-minded" who, through suffering, attain a more profound insight into the – fleeting, mortal – human condition: "morbid-mindedness ranges over the wider scale of experience, and [...] its survey is the one that overlaps" ("Sick Soul" 163). Healthy-mindedness, then, he says, "is inadequate as a philosophical doctrine, because the evil facts which it refuses positively to account for are a genuine portion of reality; and they may after all be the best key to life's significance, and possibly the only openers of our eyes to the deepest levels of truth" (163).

The purely psychological and religious implications of this conceptual opposition notwithstanding, James's identifying the "morbid" as the more valuable perspective may offer a starting point for diagnosing the direction that some of the major – avant-garde – currents of the arts have taken since the eighteenth century. These movements have shown a profound awareness of and confrontation with matters of the "morbid" kind: the demonic, evil, sickness, death, destruction, as well as various dystopian scenarios. Notably, they largely coincided with events – wars, uprisings, pandemics – that were regarded as tumultuous. It appears that the most innovative artistic vocabularies often come into being in times not of calm and prosperity, but of social or cultural distress. Perhaps it is so precisely as a reaction – or counter-reaction – to a latent belief, prevalent in such times, that the world, as it were, has lost its compass, or even exhausted its future-generating resources, and is headed towards some species of decline or incontrovertible evil. What primarily governs innovative, avant-garde vocabularies appears to be an inclination towards the negative and, often as a consequence, towards destruction. This aesthetic stance was already anticipated, in the eighteenth century, by the Romantics, whom Octavio Paz, among others, identifies as the source of the avant-gardes' tendency towards rupture – a tendency later accentuated by Impressionism in painting and Symbolism and Naturalism in literature (161). For this reason, he defines Romanticism as a separation, a splitting (183), and the modern age *per se* as the age of excision, self-negation, and criticism (210). From Romanticism, as Czesław Miłosz remarks as well, "comes the idealization of the lonely, misunderstood individual charged with a mission in society, and thus French Symbolism emerges as a specific mutation of the Romantic heritage" (368). Bourdieu likewise identifies and explains this socially charged opposition:

> Undoubtedly the hatred of the "bourgeois" and "philistines" had become a literary commonplace with the Romantics who, whether writers, artists or musicians, never stopped proclaiming their distaste for high society and the art it commissioned and consumed [...], but one cannot help noticing that during the Second Empire indignation and revolt take on an unprecedented violence, which has to be put in relation to the triumphs of the bourgeoisie and the extraordinary development of the artistic and literary bohemia. (357)

German scholar Hugo Friedrich goes to great lengths, in his *Structure of Modern Poetry*, to analyze, on a structural level, this sense of opposition in the poetry of the mid-nineteenth to the mid-twentieth century. He remarks that poetry, since Romanticism, is best defined by negative categories (19) and then predominantly discusses Symbolism's overt leaning towards the negative. In his analyses, Friedrich notices that abolition, annihilation, suspension, lacuna, emptiness, absence (127) or disorientation, incoherence, fragmentism, dislocation, estrangement (22) were some of the negative keywords of the poetry and poetics of several Symbolist authors. (These concepts, which abound in their poetry on a textual level, are also suggestively announced in titles such as the celebrated "Les fleurs du mal" [Baudelaire], "Une saison en enfer" [Rimbaud], or "Les Poètes maudits" [Verlaine].) Such negative categories, moreover, inform the aesthetic and ontological substance of Symbolist poetics; thus, Friedrich summarizes Rimbaud's aesthetics, for instance, as "an act of violence" (82, my translation) and identifies the keyword of Rimbaldian texts as cruel, "*atroce*" (82). Similarly, the essential ontological problem of Mallarmé's poetics orbits the connection between nothingness and language (125). The real and the concrete, to him, are then dislocated and sublimated into absence (126).

2 "Destruction became my Beatrice": The Attack on Language

This aesthetic leaning towards the negative is invariably followed by a fascination with destruction. This is why, for instance, Baudelaire could reflect, during the French Revolution of 1848, on the "[p]laisir naturel de la démolition" (51), or Mallarmé, invoking Dante, could declare, in a letter to Eugène Lefébure from 1867, that destruction was his Beatrice (77). The aesthetic and ontological negative, anticipated by the Romantics and Symbolists, was then further refined, on a more aggressive level, by the historical avant-gardes – which, as Ehrlicher notes, evolved primarily into a tendency towards self-destruction (35). The ontological motivation, for a generation of authors at the beginning of the twentieth century, was rooted in such disruptive events as the First World War or the Spanish Flu, which threatened to obscure the horizon of many more generations to come. As Czesław Miłosz remarks in "Ruins and Poetry," European culture, in the twentieth century, "entered a phase where the neat criteria of good and evil, of truth and falsity, disappeared; at the same time, man became a plaything of powerful collective movements expert in reversing values," and "language was appropriated by the people in power who [...] were able to change the meaning of words to suit themselves" (363). With the War, then, also came the "[m]istrust and mockery [...] directed against the whole heritage of European culture" (355). Such mistrust and mockery was expressed by the members of several avant-garde movements – most aggressively, no doubt, by the Dadaists. Thus, in the first part of the twentieth century, avant-garde authors identified with the notion of nothingness and destruction: "Every man must shout: there is great destructive, negative work to be done. To sweep, to clean" (Tzara 12).

"To sweep, to clean": The type of destruction advocated by literary and artistic avant-garde authors – primarily at the level of language, as I will expound – is, then, not destruction for the sake of destruction, but rather a gesture that aims at purging the prevailing – stagnant – vocabularies of elements deemed unnecessary, or even harmful. French-Romanian playwright Eugène Ionesco, for instance, defines the avant-gardes in terms of opposition to and rupture from established, obsolete expressions (68–69); he believes that, while most writers or artists see themselves as spokespeople of their time, the authors of the avant-gardes consciously go against their time (68). Comparing language to a fortress, Ionesco sees the avant-gardist as a critic of the present (69), an opponent of systems – an "enemy" who has broken into the fortress, and rebels against, and eventually dislocates, it from within (69). So, although the aesthetic vocabularies of the avant-gardes have often been described in negative terms, these appear nearly always invested with an innovative, liberating – even regenerative – potential.

Wittgenstein states that to "imagine a language means to imagine a form of life" (8). He defines his concept of language-games as one that "bring[s] into prominence the fact that the *speaking* of language is part of an activity, or of a form of life" (11). This definition indicates that the meanings of a word or discourse are not exclusively conveyed by the semiotic chain of signifier/signified as much as by the manner in which it is employed within a given "form of life" (which is partly analogous to the concept of *habitus* later developed by Bourdieu). In extension, then, in attacking a language – or any given semiotic system – one, in fact, attacks a "form of life," and, with it, the power relations that are inherent to that language, or to language as such. This is, I argue, what the early twentieth century avant-gardes achieved, to a degree, by way of their textual and visual experimentation and destructive practices, such as dissembling syntax, defying semantic conventions, rejecting punctuation or any form of textual coherence. It was precisely their destructive – and in part nihilistic – language games, their manifestoes, as well as their artistic-literary practices and strategies that can be regarded as a conscious effort of articulating a new "form of life," or a new vision thereof. As crucial actors of the avant-gardes proclaimed, their purpose was to attack, and thereby to change, by means of a programmatic rupture, structures widely taken for granted and regarded as commonsensical or "natural" – what Bourdieu would later define as *doxa* – within their fields and society at large. Whether or not the avant-gardists did indeed succeed in articulating a new form of (social) life remains, of course, subject to debate; yet what was primarily targeted was language – since it, to many, appeared as the root cause of humanity's failures.

Perhaps somewhat paradoxically, when faced with the urgent fear of annihilation in the aftermath of the Second World War and the ensuing atomic threat, this aesthetic leaning towards destruction, although still very much present, arguably became more tempered, so that extremely radical positions (such as the Dadaists') gradually waned. Marin Mincu notes that the "violent nihilism" of the avant-gardes in the first half of the twentieth century is no longer identifiable in the second half; a different, "recuperative attitude" can be noticed instead – one of experimentation, "in which new possibilities of expression are sought" (12, my translation). The "demolishing, undiscerning fury of the avant-gardists" is, then, replaced with "the innovating fervor of the experimentalists" (13, my translation).

Without attempting an analysis of the sociological and cultural causes of such "tempering" (which would exceed the scope of this essay), it is worth referring to Sloterdijk's remarks about nuclear fission, which, he says,

> is in any case a phenomenon that invites meditation, and even the nuclear bomb gives the philosopher the feeling of here also really touching on the nucleus of what is human. Thus, the bomb basically embodies the last, most energetic enlightener. It teaches an understanding of the essence of splitting; it makes completely clear what it means to set up a Me against a You, an Us against a Them to the point of a readiness to kill. (130)

This "understanding of the essence of splitting" – and the understanding that entire parts of humanity could potentially be extinguished in a matter, even, of minutes – may have offered a different outlook upon the notions of "evil" and "destruction." As Sloterdijk further notes, the bomb itself is "not one bit more evil than reality and not one bit more destructive than we are. It is merely our unfolding, a material representation of our essence. It is already embodied as something whole, whereas we, in relation to it, are still split" (131). The century's crises, then, gave avant-garde (and, later, experimentalist) authors an ontological justification for their leanings towards the negative and towards destruction – since, in the meantime, destruction had become the "Beatrice" of humanity as a whole.

The act, then, of attacking – even if simply at a formal or discursive level – the traditions and values that had permitted events such as the World Wars to take place was not only legitimate, but necessary. The innovative vocabularies proposed by the exponents of the classical avant-gardes and later[1] by playwrights such as Eugène Ionesco and Samuel Beckett were thus ultimately attempts at freeing the human mind from the diktat of what Ionesco defined as entrenched, obsolete vocabularies (68–69), which they sought to renew. In other words, the destruction with which the avant-gardes operated helped initiate a new type of creation, and a new conceptual and formal articulation. This arguably places them within a grey area between dystopia and utopia: Although the avant-gardists position themselves as a counter-movement against dystopian visions of decline, they do so through what may, at first glance, appear as a dystopian-inflected gesture – namely, destruction. This type of destruction, as I have tried to show here briefly, is, however, used precisely as an antidote to dystopia – namely, as the basis of utopian projects. As Ayers and Hjartarson argue, the manifestoes and programmatic writings of the avant-garde movements "presented their vision of a new life, a new society and a new man that marked a definitive break with the past, launching the readers into a utopian space of hitherto unknown life forms and experiences" (3). In times that moved dangerously on the verge of self-destruction and in which "evil" appeared, to many, to have become a genuine and highly visible "portion of reality," the negative and destruction as guiding concepts in the visual arts and literature may well have been "the best key to life's significance, and possibly the only openers of our eyes to the deepest levels of truth," as James remarks in a different context ("Sick Soul" 163).

---

[1] The dystopian events of the second half of the century coincided with the rise of a second avant-garde wave, manifesting itself primarily – and with a certain delay – in theatre: Ionesco remarks that theatre was the cultural field least marked by avant-garde experiments (73). Martin Esslin offers a possible explanation of this phenomenon by noting that "the theatre could not put [the innovations of the avant-gardes] before its wider public until these trends had had time to filter into a wider consciousness" (xii).

3 A Contemporary Example: Dys(u)topias in Forced Entertainment

As Nicole Pohl argues in this volume (27), eschatological scenarios have existed in all parts of the globe and have offered paradigmatic models for humans to be reconciled to their transitory passing through life; in this view, apocalyptic fiction may help create narratives for understanding and (re)imagining the world – even if such narratives do not always entail a redeeming quality. Perhaps the clearest expression of such an attempt to create narratives for understanding and (re)imagining the world despite – or precisely because of – the catastrophic, transformative events of the twentieth century was to be truly articulated later, by post-avant-garde – experimentalist – authors. Like their precursors, they have likewise embraced the role of "critics of the present," of "enemies" "from within the fortress," as Ionesco chose to call this positioning (69).

One such contemporary example is offered by the Sheffield-based theatre group Forced Entertainment, one of the most influential of their kind in Europe (Malzacher and Helmer 12). The core of the ensemble consists of the writer and director Tim Etchells, along with Robin Arthur, Richard Lowdon, Claire Marshall, Cathy Naden, and Terry O'Connor. As I have argued elsewhere, their theatrical vocabulary directly takes up, and transforms, several of the early twentieth century avant-gardes' practices (Tamas 116). In this sense, the theatre group has developed a deeply experimental, hybrid, post-avant-garde theatrical vocabulary that would be hardly imaginable without the innovations brought forth by the historical avant-gardes or by the likes of Beckett and Ionesco. Yet although these vocabularies are intrinsically connected, there are also a few significant differences. Forced Entertainment, for one, no longer follow the early avant-gardists' practice of manifesto-writing. Their means of expression, however, which expand the traditional boundaries of stage productions, may be seen as a new kind of manifesto. Thus, their performances, publications, web presence, along with Tim Etchells's public appearances, interviews, or art projects enhance and enlarge the group's theatrical vocabulary through a multitude of accumulated fragments, which, faithful to their artistic creed, form a meta-structure that materializes the old ideal of the *theatrum mundi*. Here, the world is re-enchanted into a space of theatrical action – and the stage, in turn, into one of reflection, but also, at times, of social and political critique. The group's explorations (as well as Etchells's art projects) range from metaphysical issues – death, madness, loneliness, or love – to environmental crises or social injustice. Just as their avant-garde predecessors, Forced Entertainment chart a historical period that has dystopian undertones – and explore, faithful to the urgent questions of their time, the workings of the new hybrid wars that now rely less on armed force than on propaganda, media manipulation, or chemical crises. At the same time, they appear to re-enchant the world and language with stories that, although not entirely utopian, still retain traces of hopefulness – and even grace. Additionally, they are armed with the "artillery" offered by cyber-space, such as their online platforms (website, social media), or DVD and YouTube recordings, which make their productions available beyond the stage. To name only one pertinent example, a recent production, *End Meeting for All* (2020), was released as a live, three-part-series of Zoom-meetings on YouTube and proposed a scenario that unfolds one year and one day after the beginning of the COVID19-induced lockdown (Tamas 121–125; Fuchs 44–48). In it, the digital becomes a stage – "a space [the performers] shared but in which [they] were nonetheless both connected and disconnected," and in which the screen was "a kind of membrane or imperfect portal between worlds" (Etchells, "Falling"). This "imperfect" – yet remarkably fecund – membrane-like quality is something most of Forced Entertainment's productions share. Quite like their avant-garde predecessors, they perpetuate a utopian-dystopian – dys(u)topian – duality in the narratives they construct, in decades that have been far from devoid of apocalyptic scenarios (or, in James's word, of a "morbid" sense of impending evil). A glimpse into two of their productions – *(Let the Water Run Its Course) to the Sea that Made the Promise* and *Emanuelle Enchanted (or a Description of This World as if It Were a Beautiful Place)* – will serve as a reflection of this.

4 *(Let the Water Run its Course) to the Sea that Made the Promise* and *Emanuelle Enchanted (or a Description of This World as if It Were a Beautiful Place)*

(Let the Water Run its Course) to the Sea that Made the Promise (1986), Forced Entertainment's fifth performance project, is "[s]et in a world that [i]s as much post-cultural as post-holocaust" (Etchells, *Certain Fragments* 134). For this production, the group received an Arts Council project grant, which facilitated the construction of their first "substantial set," made up of "two wooden floored rooms, grilled windows and a central area containing vertical pillars" (Benecke 32). This was, as Cathy Naden declared, "the first time [they] tackled urban experience and got into the whole industrial wasteland thing" (qtd. in Benecke 34). At the same time, the production marked the moment in which the group

developed some of the methods – improvisation and spontaneity in terms of content and material – that would become the hallmark of their work: "with *(Let the Water)* their own work [...] stumbled into territory that they could really call their own" (Benecke 38). As Etchells explains, the production showed "four performers enacting something that was part ritual, part game and part exorcism – a performance in which two men and two women quoted fragments of their lives, loves and possible deaths" *(Certain Fragments* 134). Their "lives, loves, and possible deaths" are enacted here in two narrative strings that intersect throughout the play: one that presents the four live performers on stage, who utter "gibberish language made up of crying, whispering, mumbling, yelling – the undocumentable shapes and architecture of language without its details" (134), and a second one that is structured as a voice-over commentary spoken by Tim Etchells and Sarah Singleton. In the first narrative string, we witness the four performers' outbursts of emotion: They tremble intensely, issue loud buzzing sounds, weep on the floor, gesture frantically, and, eventually, die – then get up, only to start again. After this visceral ritual, the partners of each couple engage in fierce fits of anger directed at each other – yet, except for the few instances in which their mumblings remotely resemble a barbaric version of English, the performers do not utter any intelligible language. The contents of their arguments remain unknown, and so does the cause – or even the authenticity, in the context of their private narratives, of their inevitable deaths. After these, intermittently, follow the weeping and mournful desperation of the "surviving" partners. As they are deprived of a comprehensible language that could, perhaps, justify their behavior, what remains for the viewer is the raw reality of anger and mourning. Human emotions, then, are here rendered absurd and, even, at times, atrocious. The scenes – with the performers' pained attempts at language, alongside a somewhat unnerving soundtrack (written by John Avery), intermittent stretches of silence, and dim lighting – endow the play with an overall dystopian, frightful undertone. Here, what Etchells terms the "undocumentable shapes and architecture of language without its details" (134) has a much less cerebral than a profoundly corporeal dimension. Siegmund remarks that several of Forced Entertainment's pieces "constantly direct the viewer's attention towards the act of speaking itself" (209). This also occurs in *(Let the Water Run its Course)*: Despite the performers' apparent inability to utter even one single coherent word, it is precisely their pained attempt at speech that draws the viewer's attention to language – either to the need for language, or, on the contrary, to its futility.

The second narrative string (the voice-over commentary spoken by Etchells and Singleton), which recurrently interrupts the first string through a sudden caesura, is more directly centered around language, as it is constituted solely by sentences uttered by the Man and the Woman. Their voices appear as negative spaces between the performers' actions, and each narrative string amplifies the occurrence and absurdity of the other. The spoken "story" can be read, to a degree, as a (jointly written) love letter, given that it starts with naming an addressee:

> Man: to the Mr Heart-Lung babies of this
> place & the so-called platitude girls
> those for whom falling is a way of life. (Etchells, *Certain Fragments* 135)

Then, later, it ends with a postscript declaimed by the Woman:

> Woman: I am writing because you asked me to write you the truth about my
> life here & because I hope we are still friends.
> I kiss you as we kissed before.
> Marina & Mr Concrete.
>
> P.S. I remember the snow, the frost, the opera building
> & your kisses.
> Isn't it funny how we never felt the cold? (139)

The letter describes someone's – perhaps their own – deaths. Yet, contrary to biological death, which occurs only once, the characters' deaths here occur gradually, step by step – death by death. The spoken story comprises five parts, each of them illustrating an individual "stage" of their collapse. This is a process that starts with an inconspicuous death, that of "his hands" – one, as the Woman declaims, that "didn't seem to matter at first, though it changed his style of dress" (135). Her opening monologue already anticipates the love story: these "wer [sic] the hands he used to 'old her hands, when they walked together in the walking hours" (135). The man's voice later closes the circle of the narrative, by identifying "part one" with the "death of her eyes" – which, he goes on, "was brilliant fun" as she was no longer able to see "the artificial sun of 1973, or the new packaging for Erotic Chocolate, or the green light by the precinct say 'Walk! Walk!' especially for her" (135). This structure runs through the first three absurdist stages – or deaths: the "death of his heartbeat" and the "death of her walk," "the death of her insides," and the "death of his skin" (136). It is only in Parts Four and Five that the narrative finally changes, so that the sections become instead "the chronology of the cities" (137) and "the love & acceptance of their true blood" (138), respectively. The chronology – an increasingly somber litany, followed, like a chant, by the Woman's line, "They spoke each night & every night about a journey standing still" (137) – comprises the "City of Stones," the "City of Wire," the "City of Rain," "the Stupid City," "the Empty City," the "City of

Variable stars," and, finally, the "City of Faith" (137). The fifth city's description calls up a dystopian image, a wasteland: "The people ther [sic] wer [sic] no lovers of water, all ther [sic] buildings wer [sic] tumbleshit, all ther [sic] trees wer [sic] bare, & the branches on the boulevards wer [sic] bones & ther [sic] wer [sic] no birds ther [sic], only bird repellent" (137).

Reality and language, here, become divided into two distinct layers: one which comprises the poetically narrating voices, and another one that creates an inarticulate idiom of grunts and silences. The love letter – like the live performance onstage – appears, to the outsider, to be mostly an absurd string of sentences, an insufficiently coherent chain of events, in part reminiscent of the Surrealist *écriture automatique*. As the dialogue and the performance advance, however, one is made to think of them as a code, as signals that the two lovers no doubt shared and recognized. In that sense, the viewer becomes a witness who pries into the lovers' intimate correspondence and into the odd mechanics of their love and deaths. From these, language and logic have escaped, and the couple's romance appears permeated with a sense of loss and a self-perpetuating chain in which love, death, and grief are indelibly tethered together.

On the other hand, *Emanuelle Enchanted* appears to create the dim possibility, or at least illusion, of utopia already through the use of the conditional in its subtitle – *as if It Were a Beautiful Place* –, while "show[ing] the events of a single night – the night of a crisis both personal and global" (Etchells, *Certain Fragments* 142). Built upon a collage-like structure, *Emanuelle Enchanted* is set in a TV newsroom with moving walls and "a panoramic glimpse of many characters presented via the cardboard signs [...] attempting visibly and not always successfully to overcome the triple hardships of theatrical representation, language and memory" (142). The short narrative fragments, spoken or read by the performers, constitute an essential part of the collage: cardboard signs, the speeches uttered in the newsrooms – "list-texts" and "longer litanic texts" (142) –, "narrational or framing texts" (142) declaimed in front of the curtain, as well as monologues. At various moments throughout the production, the performers arbitrarily pick handwritten cardboard signs off the floor and hold them against their bodies – signs that show dozens of names and nutshell-descriptions (or labels, perhaps), such as "Marcie (Pregnant)," "One Bavarian Princess," "Miss Deaf America," or "Prince Valium," to name only a few. This gesture leads to a mashup of identities, a flood of names and characters that appear to bear no connection, and no life except on the cardboard signs, since the same faces circulate these names – only for a few seconds, after which they are once again replaced. Some of the passages from the Curtain Texts are suggestive of dystopia: "In the summer when the earth changed it rained for five months and on the night the rain stopped a silence fell like we'd woke up in a silence from a dream. // We were in a city and on that strange night only the dead walked about in it, smiling and drinking halves of lager," one performer declaims (147). The speakers' descriptions of that night unfold further – their attempts to send messages to "those dead people there," who "did not fully understand our language," the "off the scale" "rhetorical temperature," the reply messages that reached them "like a FILE DELETED SYMBOL or like a drunken joke" (147). Another performer concludes, then, that the city in which they lived was one "where happy endings were not popular, and so, without pens, we preferred to write messages on the walls in our blood" – and yet, she adds, "no one knows if our messages got through" (147). This scene is then followed by Newsroom One, in which the performers read arbitrary telegram-like notes, some of which they declare to be unreadable. (Here, the stage directions dictate that the notes should be presented "like the urgent missives of a culture in the last throes of crisis" [148]). This furthermore heightens the impression that the protagonists are trapped in a dystopian space, in which telegrams reach them from a different world – as signals that they desperately try to decipher.

As I argued in the introductory sections of this essay, earlier phases of the avant-gardes (from Romanticism and Symbolism to the movements, predominantly, of the first half of the twentieth century) operated with a language of destruction, of violence – often as a counter-reaction to a latent belief that the world was on the verge of collapse. This appears, no doubt, as a dystopian-inflected "language," given its profoundly negative tenors. Oftentimes, however, it presents itself as a kind of antidote to dystopia – hence, as a foundation of utopian projects which require us to "sweep, to clean" (Tzara 12). Starting with the second half of the twentieth century, the "violent nihilism" of the earlier avant-gardes becomes more tempered, and is replaced by an attitude of experimentation (Mincu 12, my translation). The two productions I chose to exemplify this experimentalist tendency likewise deal with the duality of, and close link between, utopia and dystopia. Perhaps what takes the place of the earlier language of destruction is the delicate articulation of a "language of loss" (to borrow from Siegmund [210]). Both productions appear to engender a state that Siegmund, taking his cue from T. S. Eliot, defines as "violet hours": "times in which the world appears in a particular state of arousal, temporarily suspending the usual course of events," a "time of transition" (207–208). The moment of suspension – the in-between – in which the protagonists are caught sways adroitly between utopia and dystopia. We find them engaged in rituals, games, in which their lives and deaths permeate each other – rituals that are made visceral by their outlandish use, or deprivation, of language.

Works Cited

Ayers, David, and Benedikt Hjartarson. "New People of a New Life: Modernism, the Avant-Garde and the Aesthetics of Utopia." *Utopia: The Avant-Garde, Modernism and (Im)possible Life*. Ed. David Ayers et al. Berlin: de Gruyter, 2015. 3–13.

Baudelaire, Charles. "Mon cœur mis à nu." *Journaux intimes*. Paris: Les Éditions G. Crès, 1920. 45–112.

Benecke, Patricia. "The Making of … From the Beginnings to *Hidden J*. The Making of … Von den Anfängen bis zu *Hidden J*." *"Not Even a Game Anymore": The Theatre of Forced Entertainment / Das Theater von Forced Entertainment*. Ed. Judith Helmer and Florian Malzacher. Berlin: Alexander Verlag, 2004. 27–47.

Bourdieu, Pierre. *The Rules of Art. Genesis and Structure of the Literary Field*. Trans. Susan Emanuel. Stanford: Stanford UP, 1995.

Ehrlicher, Hanno. *Die Kunst der Zerstörung: Gewaltphantasien und Manifestationspraktiken europäischer Avantgarden*. Berlin: Akademie Verlag, 2001.

Etchells, Tim. *Certain Fragments: Contemporary Performance and Forced Entertainment*. London: Routledge, 2001.

Etchells, Tim. "Falling into Place: A Note on *End Meeting for All*." April 2020. Web. 5 Apr. 2022. <https://www.forcedentertainment.com/falling-into-place-a-note-on-end-meeting-for-all/>.

Esslin, Martin. *The Theatre of the Absurd*. Anchor Books, 1961.

Friedrich, Hugo. *Die Struktur der modernen Lyrik: Von der Mitte des neunzehnten bis zur Mitte des zwanzigsten Jahrhunderts*. Reinbeck: Rowolt, 1988.

Fuchs, Barbara. *Theatre of Lockdown: Digital and Distanced Performance in a Time of Pandemic*. London: Methuen, 2022.

Ionesco, Eugène. *Note și contranote*. Bucharest: Humanitas, 1992

James, William. "The Religion of Healthy-Mindedness." *The Varieties of Religious Experience: A Study in Human Nature. Being the Gifford Lectures on Natural Religion Delivered at Edinburgh in 1901–1902*. New York: Longmans, Green, and Co., 1902. 78–126.

James, William. "The Sick Soul." *The Varieties of Religious Experience: A Study in Human Nature. Being the Gifford Lectures on Natural Religion Delivered at Edinburgh in 1901–1902*. New York: Longmans, Green, and Co., 1902. 127–165.

Mallarmé, Stéphane. *Selected Letters*. Ed. and trans. Rosemary Lloyd. Chicago: The U of Chicago P. 1988.

Malzacher, Florian, and Judith Helmer. "Plenty of Leads to Follow: Foreword." *"Not Even a Game Anymore": The Theatre of Forced Entertainment / Das Theater von Forced Entertainment*. Ed. Judith Helmer and Florian Malzacher. Berlin: Alexander Verlag, 2004. 11–23.

Miłosz, Czesław. "Ruins and Poetry." *To Begin Where I Am: Selected Essays*. Ed. Bogdana Carpenter and Madeline G. Levine. New York: Farrar, Straus and Giroux, 2001. 352–370.

Mincu, Marin. *Eseu despre textul poetic*. București: Cartea Românească, 1986.

Paz, Octavio. *Los hijos del limo: Del romanticism a la vanguardia*. Barcelona: Biblioteca de Bolsillo, 1989.

Siegmund, Gerald. "The Dusk of Language: The Violet Hour in the Theatre of Forced Entertainment / Die Abenddämmerung der Sprache: Die blaue Stunde im Theater von Forced Entertainment." *"Not Even a Game Anymore": The Theatre of Forced*

Entertainment / Das Theater von Forced Entertainment. Ed. Judith Helmer and Florian Malzacher. Berlin: Alexander Verlag, 2004. 207–219.

Sloterdijk, Peter. *Critique of Cynical Reason*. Minneapolis: U of Minnesota P. 2001.

Tamas, Luciana. "End Meeting for All: The Performative Meta-Collages of Forced Entertainment." *Journal of Contemporary Drama in English* 9.1 (2021): 114–127.

Tzara, Tristan. *Seven Dada Manifestoes and Lampisteries*. London: Calder Publications, 1992.

Wittgenstein, Ludwig. *Philosophical Investigations*. Oxford: Basil Blackwell, 1986.

Sebastian Berg
The End of Capitalism and the End of Democracy: Dystopian and Critical Utopian Political Economies in an Age of Austerity

In 2019, Wolfgang Streeck, a German political economist and until recently head of the Max-Planck-Institute for the Study of Societies in Cologne, commented on Britain's recent past:

> Under the Blair, Brown and Cameron governments, the UK turned into the training ground and experimental arena par excellence of a liberated, fanatical privatization, into the dystopia of a deliberate incorporation of the foundational into the profit economy with disastrous consequences for both civic equality and economic efficiency. ("Kommunismus" 96, my translation)

Critiques of neoliberalism like this one have become more frequent over the last fifteen years. Still, it is surprising that, with Streeck, it is voiced by someone who, twenty years earlier, had urged the German government to consider the following: "State and public administration must be accompanied by the dynamic of the market – not least to breathe new life into fossilized bureaucracies, petering out programmes and frustrated individuals." ("Arbeit," my translation.) At that point, Streeck identified Oscar Lafontaine, who had recently resigned as financial secretary and leader of the Social Democratic Party, as the main obstacle to modernisation (since he had advocated a "vulgar Keynesianism") and the North West European countries, especially the UK and the Netherlands, as role models for welfare state and labour market reforms in Germany. Streeck argued that wages in Germany's embryonic service economy were too high, job seekers too inflexible, and thus official and hidden unemployment figures much higher than in less regulated, more liberal capitalist economies. Self-critically, he commented on his former positions a couple of years ago:

> We hoped we could ride the wave of marketisation to save the welfare state by making it fit for a global economy. [...] It soon became clear that this was the last round for European social democracy and that we were not winning. We were reforming capitalism, and only later noticed that capitalism had been reforming under our very eyes. ("Last Rounds")

Note: This text is written in memory of two "Milibandisti" whom I met on a number of occasions and who were important to me (as to many others) – as persons and as sources of intellectual inspiration: David Coates (1946–2018) and Leo V. Panitch (1945–2020).

Following this epiphany, Streeck adopted Marxist and Polanyian notions of capitalism and started publishing his reflections in left-wing outlets such as *Blätter für deutsche und internationale Politik* in Germany, *Jacobin* in the USA, but especially *New Left Review* and the publisher linked to it, Verso, in Britain. Hence, he has started playing a visible role in current left-wing academic-activist discourse in Britain and the Anglophone world.

However, he is not the only critic of neoliberal capitalism and austerity. My contribution reads Streeck's work in conjunction with that of another political economist of some standing in the Anglophone left, David Coates, British by birth, US citizen, and until his death in 2018 Worrell Professor of Anglo-American Studies at Wake Forest University, North Carolina. Like Streeck, Coates modified his perspectives on capitalism and neoliberalism over the last fifteen to twenty years, though far less fundamentally. In 2005, he applauded Gordon Brown's (then Britain's Chancellor of the Exchequer and according to Coates and others, the chief architect of the New Labour government's economic policies) plans to eliminate child poverty and congratulated him on his "progressive heart" (*Prolonged Labour* 79). At the time, Coates granted Brown the benefit of the doubt, though he held serious reservations about the New Labour project:

> This is a Chancellor, after all, who has played an important role on the world stage, seeking to alleviate the burden of debt on the most oppressed and underdeveloped of Third World countries. It is also a Chancellor who, with his prime minister, has publicly committed this government to the halving of child poverty in one decade and to its removal in two. It is less the radicalism of his aims than the moderate nature of his chosen means that has raised, and continues to raise, questions on where Gordon Brown sits on any conventional left-right spectrum of British politics. (*Prolonged Labour* 59)

More recently, however, Coates described the New Labour project in a more condemning mood as "Thatcherism-lite" (*Flawed Capitalism* 86), or "neoliberalism with a softer face" (*Flawed Capitalism* 87), and argued, with regard to the financial sector's dominance in the British economy, that "[i]f Labour had spent its years in power correcting that imbalance, and retooling UK industry for the new competition, it would likely still be in government and the UK still in the EU" (*Flawed Capitalism* 187). However, there is a fundamental difference between both scholars' positions: whereas Streeck continuously voices his conviction that capitalism will collapse under its own contradictions and will give way to what he calls an "interregnum" (quite close to what others have called barbarism), Coates retains some hope that capitalism can be re- and transformed in a way that avoids breakdown of the kind Streeck anticipates (barbarism's alternative, a form of socialism, remains possible).

Streeck and Coates make an interesting comparison because both are political economists, born in the same year (1946), acting to some extent as public intellectuals, and writing for left-wing political-academic publications – while Streeck regularly contributes to *New Left Review*, Coates served as one of the most active contributing editors to *Socialist Register*, an annual founded as a response (but one of critical appreciation) to *New Left Review*. On the other hand, Streeck was a card-carrying social democrat for a long time (both figuratively and literally) and only later took a more radically critical position towards capitalism, whereas Coates started as a Trotskyist who modified his perspectives and became one of the leading "Milibandisti."[1] Analysing their work, my contribution asks about the reason why the long-term critic of capitalism (Coates) seems to be more optimistic today than the social democrat turned pessimist (Streeck) that capitalism can be reformed or transformed in ways that makes it compatible with democracy. I am less interested here in Streeck and Coates as *persons* but as *examples* of certain discursive positions in the political left on capitalism's current state and likely future. One position maintains the critically-utopian conviction that another (better) world is possible, while the other retreats to dystopian fatalism. Avoiding vulgar psychology, I intend to explain the differences observed from within the internal coherence of the perspectives taken. It seems to me that

1. Streeck tries to formulate a general critique of *capitalism* as an institutionalised system, whereas Coates attempts a critique of *capitalisms* as outcomes of specific power configurations and struggles;
2. Streeck follows politically trendy hegemonic imaginaries of contemporary capitalism as they have been popularized over the last couple of decades, while Coates has remained more immune to these shifting perceptions.

My argument is that (1.) in order to identify possibilities of political agency, one has to combine Streeck's more general with Coates' more specific critiques of capitalism, and that (2.) hegemonic imaginaries of capitalism need to be ques-

---

[1] The characteristic feature of the Milibandisti is their analysis of the resources and limits of labour movement agency in Britain, which is for better or worse closely linked to the Labour Party. David Coates and Leo Panitch defined the core of Ralf Miliband's perspective on the Labour Party and its role in British politics as the party's over-reliance on parliamentarism, its contributions to the stabilisation of British capitalism and its occasional rhetorical commitments to radical policies when in opposition which prevented the development of a more genuinely socialist party in Britain (72). However, the Milibandisti never "gave up" on the Labour Party – rather, they tried to analyse the preconditions for a sustainable radicalisation of the party which they saw as the most realistic way forward for political transformation in Britain.

tioned. The following analysis discusses both writers' reflections on capitalism, in particular its Keynesian-welfarist and Hayekian-neoliberal varieties, their effects on social formations, possibilities, limits and resources of political agency, and the role to be played by centre-left parties. The contribution finally returns to the question of how to evaluate the future of capitalism and democracy.

Capitalism

In their most recent book-length publications, both Streeck (*Capitalism*) and Coates (*Flawed Capitalism*) contribute to a debate among social scientists on likely scenarios for the future of capitalism. There is a lot they agree on. Both present rankings of varieties of capitalism. Streeck primarily contrasts the superior postwar welfare capitalism of the three decades after the end of World War II with the inferior neoliberal variety that succeeded it (*Capitalism* 4). For Coates, this historical distinction is as central as for Streeck; however, he adds another spatial differentiation: he highlights the specific deficiencies of a "liberal market capitalist economy" in comparison to a "coordinated market capitalist economy" (*Flawed Capitalism* 18). He locates the former in the USA and the UK and calls it "flawed capitalism" – admitting that all capitalisms are flawed, first of all by the capital-labour contradiction, but contending that some are more seriously flawed than others (*Flawed Capitalism* 18).

Post-war welfare capitalism in both its Anglo-American and Continental European varieties relied on a number of preconditions. First of all, the relative balance of power between capital and an organised working class forced the former into what Streeck calls a "shotgun marriage" with democracy (*Capitalism* 20). In the postwar era, capital was inclined to compromise with the working class for a couple of reasons. The history of the first half of the twentieth century had shown the disastrous consequences of an uncoordinated capitalism that had caused depression and the rise of fascism. The post-war conjuncture offered prospects of a virtuous circle of relatively wide profit margins for capital through producing for higher living standards for working-class people, creating an affluent society with high levels of aggregate demand. Industry produced for national markets and profits were reinvested in national economies. Governments played an important role in these economies, using redistributive and administrative measures of regulation. They pursued Keynesian fiscal and financial policies, transferred sectors of the economy into public ownership, engaged in macro-economic planning. The specific mix differed in individual countries. This framework, the golden age of social democracy, the *trente glorieuses*, the post-war consensus, remained in place until the 1970s. Then it disintegrated. The reasons were manifold. Streeck emphasises that profits and thus distributional margins had narrowed over time and that the "profit-dependent classes" thus looked for alternatives to the marriage of capital and nation-state-based democracy, especially in the aftermath of the oil crisis and the unravelling of the Bretton Woods agreement (*Capitalism* 16). He lists a number of changes that began in the 1970s: international economic cooperation was intensified via trade agreements. The EEC and other transnational institutions started moving towards common market policies. The communications revolution allowed for new channels of collaboration. The combination of these innovations became known as globalisation (*Capitalism* 22).

With globalisation, as both authors agree, the role of the states changed. Their economic policies switched from demand to supply management and they competed to become attractive for foreign direct investment. Coates speaks of a

> new growth theory then becoming fashionable in policy-making circles – post neoclassical endogenous growth theory – the one that prioritized the transformation of the existing welfare state into a social investment state, and the transformation of education policy from a tool of social reform into one for strengthening human capital. (*Flawed Capitalism* 87)

In practice, faced with increasing levels of redistributive conflict, governments turned to financing social policy via debt and transmuted, to use Streeck's terminology, from "tax states" to "debt states" (*Capitalism* 16). In his monograph *Buying Time*, Streeck shows that this became possible through the liberalisation of the financial sector. Speculation in financial markets created money which could be lent to states. Through these transactions, the financial sector, however, became ever more powerful. In the 1990s it had reached a state of maturity which allowed it to start demanding the servicing of debt and thus ushered in the next state project – that of the "consolidation state" (*Capitalism* 16). Since state finances had to be balanced, further chunks of state services were privatised. As a consequence, people who wanted to use them had to pay fees and ran into private debt. For welfare state societies, this meant what Streeck's colleague and occasional collaborator, Colin Crouch, called "privatized Keynesianism," and for the European Union project the transformation of an embryonic social Europe into, to use Streeck's words, a "liberalization machine" producing the "Hayekisation of EU capitalism" (*Buying Time* 114). In 2014, Streeck hence self-critically remarked:

> As the sensible social democrat that I have long been, I concede with shocked astonishment that the really important questions today are those most likely to be discussed in the vicinity of movements like ATTAC: questions about how globalisation might be retai-

lored or even – *horribile dictum* – scaled back to become compatible with egalitarian democracy. ("Nostalgia" 218)

With the rise of the debt state, a second sovereign developed beside the "state people." Streeck calls this second sovereign the "market people" (*Capitalism* 24). Credit ratings became more important for governments than polls and "the clock [was] ticking for democracy" (*Buying Time* 5). For Streeck this constitutes a hermetic dystopian development since, in the long run, disorganised capitalism is going to disorganise itself because it cannot exist without institutional regulation. This is why Streeck predicts an "interregnum" (*Capitalism* 36), characterised by social entropy – the development of an under-institutionalised society that will succeed capitalism as it is currently known.

In several respects, Coates tells a very similar story: the two post-war projects of stabilising capitalism eventually failed – the Keynesian one in the 1970s, the neoliberal one in the 2000s (*Flawed Capitalism* 152). He also agrees with Streeck that failure has to be related to an increasingly unequal distribution of power between the forces of capital and labour (*Flawed Capitalism* 258). There is, however, an important difference: for Coates, this state of affairs is not an unavoidable development but the consequence of wrong political choices. He shows this by pointing to the variety of governmental reactions to the developments sketched out by Streeck and argues that, in the British case, not only governments of the right but also the centre-left relied on excessive private credit debt, low wages and a small low-productivity manufacturing sector (*Flawed Capitalism* 88–89). Governments elsewhere, for example the Swedish one, chose a different strategy with higher wages, higher investment and higher inflation (Coates, "Labour Power"). The British strategy went hand-in-hand with an increasing distance that the British Labour Party tried to put between itself and the traditional labour movement, especially the trade unions. Coates does not understand recent developments as capital's dystopian takeover of government but as a concerted strategy pursued by both. This still leaves space for an alternative government with an alternative strategy – if it manages to gain the support of a sufficient number of people in society.

Social Formation

Streeck's and Coates' views on the state of society in contemporary neoliberal capitalism differ. For Streeck, the social formation has to a large extent disintegrated. He identifies two class-like groups, of which only one understands itself as a class. This is the group of the super-rich. They have reached a level of independence from society that allows them to care about themselves only – which might include engaging in donating and philanthropic activity. Donations buy political majorities (Streeck in particular refers to the USA here). Philanthropism buys public legitimacy; furthermore, it aims at depoliticising social inequality and restricting welfare to acts of altruism. For people beyond the group of the super-rich, the "age of entropy" simply "de-socialises" and "de-institutionalises" society into a "post-social" one (*Capitalism* 13). Hence most people rely for their survival on their permanent self-perfection and the adaptation of their marketable skills. They are confronted with "disruptions," new directions that capitalist development takes, with new demands, to which they have to react creatively. Ideally, they become "resilient" to such disruptions. Streeck observes four resilience strategies, all of them to be performed by individuals rather than collectives: coping, hoping, doping, and shopping (*Capitalism* 41–44). Coping simply implies relying on one's individual stamina and creativity to deal with adverse situations, as they are characteristic of a de-institutionalised society. Hoping supports coping in that one imagines a better life waiting in the near future as a reward for successful coping. Doping helps both coping and hoping: one can facilitate coping via the performance-enhancing use of all kinds of potential "substances" and other distractions in the widest sense (from cocaine to fitness studios), but one can also alleviate hoping through similar commodities (from alcohol to packaged holidays) – as escapist means that seemingly (and temporarily) relieve one from the pressures of disruption and resilience. Finally, shopping serves the satisfaction of one's desires beyond human needs which are relentlessly advertised as "dope": as rewards for hard work and for proving resilience in the face of disruptions. With the provision of "dope" and the creation of desires, the economic cycle is completed because production and financial services need a reliable market in which people buy, borrow money, and to whose production processes they sell their labour power. As a consequence, individuals embrace a culture of a neo-Protestant work ethic combined with a competitive hedonism (*Capitalism* 45). As a corollary, all non-consumerist gratifications are discredited and all collective efforts of breaking this cycle seem futile and are interpreted as symptoms of personal weakness (*Capitalism* 45).

Although Coates does not completely disagree with Streeck, he adds nuances and adopts a slightly different perspective. As mentioned before, for him, not capitalism alone produces a certain type of society, but political agents are also involved that frame, perform and popularise both general ideas of society and concrete social practices. Referring to his colleague Colin Hay, Coates points out that it is one of the core characteristics of neoliberal policies, embraced and propagated by both right-wing and centre-left parties, that (at least until recently) they have not presented any theory of society beyond an image of a social formation composed of supposedly rational individuals who are – here his view is quite similar to Streeck's – increasingly exposed to risks and forced to react to them. However, whereas Streeck seems to identify the people as forming a more or less monolithic albeit amorphous mass (apart from the super-rich), Coates presents a more detailed analysis of the situation of different groups and classes within society. He shows that working-class people (especially in the old industrial, pro-Brexit areas), black, Asian and minority-ethnic (BAME) people, and young workers and professionals are particularly threatened by risks (*Flawed Capitalism* 186). He also sees women and single-parent families being forced into the high-risk group. Their responses, according to him, are to some degree individual coping (and maybe hoping, doping, and shopping) strategies; however, this is not the end of the story. Over the last couple of years, many collective and explicitly political reactions could be observed. Coates does not approve of all of them; for example, he interprets the election of Trump (which he deplores) and the majority vote for Brexit (which he regrets) as acts of political protest – people opted for the demagogue and for nationalism because, as he succinctly put it, they had lost confidence in the future (*Flawed Capitalism* 184). Other political reactions he evaluates positively. Observing omnipresent signs of stress among the generation of millennials, he sees them moving politically to the left – this is corroborated by the fact that more millennials voted for Sanders than for Clinton and Trump together in 2016 and by this generation's equally strong support for Corbynism (*Flawed Capitalism* 273). For Coates, in other words, the future fates of capitalism and democracy do not simply follow a predetermined path of entropy. Once more, there is still space for politics – and dystopian or utopian futures depend on political agency, including that of parties, trade unions and other political organisations and movements.

Political Agency

Streeck observes that traditional forms of political agency have eroded over the last decades. While the tax state, especially once it had run into crisis in the 1970s, was characterised by high levels of strike action, the debt state restricted agency to the electoral arena – Margaret Thatcher's successive laws to curb trade union activity and the showdown with the miners immediately spring to mind here. The consolidation state, finally, its hands tied by debt service, had no choice but to put most political decisions into the hands of an expertocracy of international organisations (for example, G7), councils of ministers (for example, at EU level) and central bank executives (for example, the ECB in the case of the European Monetary Union area). In other words, government has been transferred to the "postmodern clientelism of the new financial capitalism" ("Nostalgia" 217). The new sovereign (the market people) has clearly won the fight against the former sovereign (the state people), and a reversal is either impossible or unlikely because

> the fragmentation of the working class and the debasement of work through excessive commodification and flexibilization preclude the formulation of a coherent oppositional project, like socialism, aspiring to separate what is progressive from what is reactionary in capitalism and preserve it (*Capitalism* 27).

Instead, party membership is in decline, party systems are fragmenting, voting becomes more unreliable, membership in trade unions shrinks, and people turn to "populist" movements and their leaders in increasing numbers – those that distinguish good people and corrupt elites in an over-simplifying way.[2] If there is a chance remaining for democratic renewal, however remote, it lies with the political institutions of the nation states, which at least in some cases worked as a rear-guard, slowing down the erosion of democracy: "[I]f we did not have them [...] the project of a democratic political economy would not be as troubled as it is today but it would long have ceased to exist." ("Nostalgia" 219) Still, there is neither a guarantee of nation-state institutions performing this role nor a correlation between a state's power and its commitment to defending democracy, a thesis that Streeck sees vindicated by a prominent example: "And if one wants to see how even a large state can appeal to market constraints to dispossess democratically and economically the ordinary-people majority of its population, one need look no further than the United States." ("Nostalgia" 218)

While Streeck sees the hijacking of governments by the market people as the unavoidable consequence of the disempowered consolidation state, Coates suspects this claim to be an ideological smokescreen. Convinced that state debt is much less serious than private debt and trade imbalance, he again accuses most administrations of simply having taken the wrong decisions. He condemns the New Labour governments' "Faustian pact" with the UK's financial institutions, revealed for example, by granting independence to the Bank of England (*Flawed Capitalism* 89), which led to the banking and debt crisis, and the coali-

---

2 Populism became widely discussed in the social sciences over the last decade. The most generally shared conceptualisation of populism is as an ideology that contrasts a monolithic honest people with a corrupt elite. The problem with the populism concept is its lack of differentiation between left-wing and right-wing populisms (see Berg "Populism" and "Against 'Populism'").

tion government's reckless redistributive policies to deal with its consequences. The latter were disastrous for two reasons:

> One is the disproportionate burden that his [George Osborne's, chancellor 2010–2016] policies placed on those least able to bear them. The other is that all this pain – voluntarily inflicted on others as it was – did not in the end fundamentally transform the UK's ongoing economic weaknesses. (*Flawed Capitalism* 232)

It is this experience that drove people towards a right-wing populism which projects these mistakes onto institutional arrangements beyond the nation state. This populism "balks at free trade, opposes the free movement of people, and looks to strengthen the national economy pulling back (or in the UK case, entirely breaking from) [sic] existing international agreements and institutions" (*Flawed Capitalism* 254). Left-wing political agency, according to Coates, should thus take the nation state seriously but not in the way propagated by right-wing populists. The left should "directly challenge existing patterns of power and reward" at the nation-state level but at the same time be aware of the state's embeddedness in wider economic, social, and environmental contexts, which influence its room for manoeuvre (*Flawed Capitalism* 252). In this dual process of action and reflection, centre-left political parties still have an important role to play.

Centre-left Parties

Both writers agree that centre-left parties embraced the neoliberal globalisation project too enthusiastically and redefined social policy as public provision of private competitiveness in the global market (Streeck, *Capitalism* 22). As seen in the introduction, Streeck critically reflects on his own role in this operation, while Coates, involved in the *Socialist Register* project, which started evaluating globalisation critically in the early 1990s, had always been more sceptical. After the crisis of 2008, Streeck became a scathing critic of both the EU and European monetary union and accused centre-left parties of embracing the "European project" either naively or strategically. He argues that behind their proclaimed progressive and transnational Europeanism both the German Social Democrats and Germany's largest industrial union (the *IG Metall*) continuously prioritised German manufacturing interests (jointly with the Christian Democrats and the industrial employers' association). In particular, he condemns monetary policies that rule out any possibility of devaluing a country's currency (pressing the weaker national economies into the role of import markets for the industrial output of the stronger ones), while at the same time forcing countries to accept unlimited migration (transforming the surplus labour of the weaker national economies into a reverse army of workers in the stronger ones). What centre-left politicians advertise as liberal cosmopolitanism, is in his eyes nothing but the means to inflate the labour supply with the goal of driving down wages in the centres of production. Thus, Streeck sees no way forward for the centre-left unless it re-evaluates its relationship with the nation state, since in recent years states were the only institutions which "to some extent managed to oppose techno- and moneycrats" ("Nostalgia" 218). Hence for him, centre-left policies should defend institutions at the national level, attacked by a "neoliberal-supranational Leviathan," by "converting the remains of postwar social democracy into barricades against technocratic encroachment" ("Nostalgia" 219).

Again, Coates goes along with several aspects of Streeck's analysis. By accepting large parts of the Thatcherite settlement, the Labour Party abandoned its ideological core, labourism (often criticised for its lack of transformative ambitions by left-wing critics like Coates in the past), and political purpose – to act as a vehicle for the realisation of working-class aspirations. He particularly deplores the severing of the union-party link under New Labour. What is required is a fundamental break with New Labour's third way,[3] a break that he sees as being implemented by the party under the Corbyn-McDonnell leadership. A new counterhegemonic project, promising a new social settlement, is needed, seems within reach, and proves to be popular, gaining the support of about 40 per cent of the electorate in the 2017 elections. The task he identifies sounds quite similar to the 2017 slogan ("for the many, not the few"):[4] "[T]he job of the left now is to move us away from a capitalism flawed by privilege into one made less privileged by the assertion of popular control." (*Flawed Capitalism* 21) Coates presents a concrete list of top priorities which moves beyond and is more detailed than Streeck's defence of the nation state against supranationalism – he suggests a new social structure of accumulation, redirecting military production and financial services to a "civilian-focused industrial and manufacturing renewal" (*Flawed Capitalism* 263), extending public ownership, reintroducing more progressive taxation and industrial democracy, strengthening the

---

[3] The concept of the third way was formulated by Anthony Giddens, who proposed it as an alternative to traditional left-wing and right-wing positions, and adopted by the Labour Party under Blair to distinguish its approach to social, economic and tax policy from both "Old Labour's" welfarism and Thatcherite neoliberalism.

[4] The formulation of organising economic life in a way that creates benefits and wealth "for the many not the few" was, ironically, part of the new Clause IV which New Labour substituted for the original demand for the socialisation of the means of production.

influence of trade unions (*Flawed Capitalism* 266). He proposes using environmental concerns "as a beachhead to establish the legitimacy of the more general regulation of private industry by the democratic state" (*Flawed Capitalism* 266). Coates points to the existence and recommends the study of innovative and detailed proposals for economically interventionist states.[5] Chances of realising such a programme depend on simultaneous changes at the international level – especially new trade rules (terminating blindness towards social and ecological "externalities") and capital controls (curbing speculation and tax evasion). Anticipating strong resistance, Coates leaves no doubt that centre-left parties must liaise with, and integrate themselves into, a wider movement for social justice (*Flawed Capitalism* 276). In particular, he sees a need to listen to, and take on board, the young generation of the millennials. Once elected into power, governments would have to act fast. The task is enormous, but Coates sees this as a challenge rather than an excuse for restricting centre-left activism exclusively to the defence of the nation-state institutions as they exist today – he quotes the former leader of the Independent Labour Party, Jimmy Maxton: "[I]f you can't ride two horses at once, you shouldn't be in the circus!" (*Flawed Capitalism* 252)

Against Fatalism

Streeck's and Coates' analyses obviously overlap in many respects. Still, they also reveal important differences. Streeck's narrative of the changes of capitalism takes an institutionalist perspective. Institutions are understood as embodied, corporate entities: they act, they develop a life of their own. They stabilise, become firmly entrenched and almost immune to attempts at challenging and changing them. Capitalism has become a historical social order to stay until it destroys itself (Roos 254, Klein 93). According to Streeck, there is no emancipatory project in sight to challenge it seriously. Jerome Roos suspects that this view is simply the inverted logic of Streeck's earlier position:

> [A]fter the crisis, social-democratic optimism becomes post-democratic pessimism; "wishful thinking" about the compatibility of capitalism and democracy a dogged insistence on their inherent incompatibility; the unquestioning faith in gradual progress becomes the absolute certainty of irreversible decline. (Roos 283)

---

[5] He refers to the works of Mariana Mazzucato on the entrepreneurial state and Colin Hay and Tony Payne on civic capitalism.

Coates' account, on the other hand, is structural rather than institutionalist: it pays closer attention to agents and dynamics – he describes a dialectical relationship between institutions, historical developments and individual and collective agency. His perspective aims at identifying actors and levers for achieving political change. In particular, he does not write off labour movement organisations – which might be shadows of their former selves but still exist and thus could be revitalised, albeit in modified forms.

In his writings Streeck continues to adopt imaginaries of neoliberalism that have been circulated and become hegemonic at different times. His proposals from the late 1990s testify to his acceptance of the at the time widely-held trickle-down theory – everyone has the chance to profit from neoliberal globalisation to some extent as long as nation-state governments create the framework and infrastructure enabling their countries to excel in global competition. Since 2008, however, a new imaginary has been substituted for the previous one by Streeck and others: capitalism is disastrous and either invincible or self-destructive. The only chance for individuals (or collectives of all sorts) consists in becoming resilient to its dangers and survive – in competition with others over increasingly scarce resources and in (pre-emptive) self-defence against likely catastrophes. Ideas of collective projects and (re-)actions are mostly absent from Streeck's analyses; it is only recently that he has started identifying the decommodification of the foundational economy at local level as a likely source of an "everyday communism." However, a critical political economy perspective moving beyond dystopian fatalism should consider much more – from philosophical reflections on alternative hedonisms (Soper) to practical experiments with community wealth building (Brown and Jones). As sociologist Luke Martell pointed out, the policies are there, it is the politics that is missing. Perhaps this politics, the imaginary of a fully-fledged counterhegemonic project that Coates urges the left to embrace, should be called socialism (a term that Streeck obviously meticulously avoids and which only comes up rarely in Coates' last book as well). Definitely such an imaginary should consider the passage Coates' book ends with: "[I]t is now time for progressives to reassert their commitment and confidence [...] and to ask the ultimate political question: 'Who is to be the master here?' Let us hope, for the good of our children and grandchildren, that ultimately it will be them." (*Flawed Capitalism* 276)

Works Cited

Berg, Sebastian. "Populism in the Social Sciences." *Hard Times: Deutsch-englische Zeitschrift* 101 (2018): 12–14.
Berg, Sebastian. "Against 'Populism': Protest and Power in a Postpolitical Age." *Crisis, Risks and New Regionalisms in Europe: Emergency Diasporas and Borderlands*. Ed. Cecile Sandten et al. Trier: Wissenschaftlicher Verlag, 2017. 47–60.
Brown, Matthew, and Rhian E. Jones. *Paint Your Town Red: How Preston Took Back Control and Your Town Can too*. London: Repeater books, 2021.
Coates, David. *Flawed Capitalism: The Anglo-American Condition and Its Resolution*. Newcastle: Agenda Publishing, 2018.
Coates, David. *Prolonged Labour: The Slow Birth of New Labour Britain*. Basingstoke: New York: Palgrave Macmillan 2005.
Coates, David. "Labour Power and International Competitiveness: A Critique of Ruling Orthodoxies." *Global Capitalism versus Democracy*. Ed. Leo Panitch, Colin Leys. London: Merlin Press 1999, 108–141.
Coates, David, and Leo Panitch. "The Continuing Relevance of the Milibandian Perspective." *Interpreting the Labour Party: Approaches to Labour Politics and History*. Ed. John Callaghan, Steven Fielding, and Steve Ludlam. Manchester: Manchester UP, 2003. 71–85.
Crouch, Colin. "Privatised Keynesianism: An Unacknowledged Policy Regime." *British Journal of Politics and International Relations* 11.3 (2009): 382–399.
Giddens, Anthony. *The Third Way: The Renewal of Social Democracy*. Cambridge: Polity Press, 1998.
Hay, Colin, and Tony Payne. *Civic Capitalism*. Cambridge: Polity Press, 2015.
Klein, Dieter. *Zukunft oder Ende des Kapitalismus? Eine kritische Diskursanalyse in turbulenten Zeiten*. Hamburg: VSA, 2019.
Labour Party. *The Labour Party Manifesto: For the Many not the Few*. Northumberland: Potts Print, 2017.
Martell, Luke. "What is Starmerism? – Reflections on Some Early Indications." *Labour Hub*, 9 May 2020. Web. 30 Dec. 2021. <https://labourhub.org.uk/2020/05/09/what-is-starmerism-reflections-on-some-early-indications/>.
Mazzucato, Mariana. *The Entrepreneurial State: Debunking Public vs. Private Sector Myths*. London: Anthem, 2014.
Roos, Jerome. "From the Demise of Social Democracy to the 'End of Capitalism.'" *Historical Materialism* 27.2 (2019): 248–288.
Soper, Kate. *Post-Growth Living: For an Alternative Hedonism*. London: Verso, 2020.
Streeck, Wolfgang. "Der alltägliche Kommunismus: Eine neue Ökonomie für eine neue Linke." *Blätter für deutsche und internationale Politik* 6 (2019): 93–105.
Streeck, Wolfgang. *How Will Capitalism End? Essays on a Failing System*. London: Verso, 2017.
Streeck, Wolfgang. "Social Democracy's Last Rounds." *Jacobin* (2016). Web. 30 Dec. 2021. <https://www.jacobinmag.com/2016/02/wolfgang-streeck-europe-eurozone-austerity-neoliberalism-social-democracy/>.
Streeck, Wolfgang. *Buying Tim: The Delayed Crisis of Democratic Capitalism*. London: Verso, 2014.

Streeck, Wolfgang. "Small-State Nostalgia? The Currency Union, Germany, and Europe: A Reply to Jürgen Habermas." *Constellations* 21.2 (2014): 213–221.
Streeck, Wolfgang. "An Arbeit fehlt es nicht". *Der Spiegel* 19 (1999). Web. 30 Dec. 2021. <https://www.spiegel.de/politik/an-arbeit-fehlt-es-nicht-a-23336b01-0002-0001-0000-000013220370>.

Dennis Henneböhl
Utopian Past and Dystopian Present? Nostalgia in Brexit Britain

In times of austerity politics, deep divisions over political issues such as Brexit, and, more recently, a global pandemic, utopian wishes for a better place abound in Britain. Indeed, according to Fátima Vieira, utopian thought can be defined as "the desire for a better life, caused by a discontentment towards the society one lives in [...]. Utopia is then to be seen as a matter of attitude, as a kind of reaction to an undesirable present and an aspiration to overcome all difficulties by the imagination of possible alternatives" (6–7). Whereas the utopian world is nowadays most often projected into the future, especially due to the growing overlap with the science-fiction genre, this has not always been the case. As Vieira points out, this shift occurred only during the late eighteenth and early nineteenth centuries (11). When Thomas More first coined the term, utopias such as his eponymous island were firmly located in the present, more specifically on remote islands or in unknown lands (Vieira 9). Moreover, the existence of utopian thought itself long predates that of More's term. In fact, it can be traced back as least as far as Antiquity (Vieira 5). Here, the utopian alternative is rather located in the past, more specifically during a lost "Golden Age" (Vieira 5). This notion connects to another way in which people react to a negatively perceived present, namely nostalgia. According to Stuart Tannock, nostalgia "invokes a positively evaluated past world in response to a deficient present world. The nostalgic subject turns to the past to find/construct sources of identity, agency or community, that are felt to be lacking, blocked, subverted or threatened in the present" (454). John Storey similarly compares nostalgia to utopian thought (107). He remarks upon their similarities in terms of the language used to define these two concepts as well as their critical focus on the present. They only differ in the fact that nostalgia locates the better place exclusively in the past. In order to describe this specific form of utopian thought, Storey coined the term "utopian nostalgia." Here, "[t]he past, like the future or elsewhere [in more general utopian thought], is used to articulate the uncertainties and disruptions of the here and now. In other words, appropriate the past in order to reorganize the present" (107). Thus, this kind of nostalgia constructs a narrative about the past as a utopian Golden Age.

One area where utopian nostalgia can be identified is contemporary political rhetoric. In fact, Storey developed this concept to describe the 2016 election campaign of Donald Trump (107–113) and in the following paragraphs it will be shown that utopian nostalgia can also be applied to the issue of Brexit. This chapter offers a case study of nostalgic rhetoric in contemporary Britain not only with regard to the 2016 referendum campaign and subsequent negotiations with the EU over the country's post-Brexit future, but also to the more recent event of the COVID-19 pandemic, where nostalgia is instrumentalised in a similar manner. The selection of primary texts mainly focuses on the leading figureheads of the Leave campaign like Boris Johnson as well as politicians who initially supported Remain like Theresa May. The material includes political speeches, interviews and comment pieces published in national newspapers as well as statements made in traditional as well as social media. My analysis combines theories from British Cultural Studies like national identity with methods from Rhetoric Studies such as Aristotle's three artistic proofs of *logos*, *ethos*, and *pathos*, which "[have] generally been accepted by classical scholars and [are] still considered relevant in understanding persuasive language and rhetoric" (Charteris-Black, *Analysing* 8).[1] More importantly, I will employ a narrative approach to both political rhetoric as well as nostalgia. Whereas "narrative criticism" has long been an established school of Rhetorical Studies (e.g. Foss 307–319, Rowland 125–145), there is as yet no comprehensive framework for a narrative approach to nostalgia. Thus, I will be drawing on various theories and concepts about narratives and nostalgia in a first step towards creating such a framework.

As Michael Müller and Petra Grimm point out, all narratives consist of at least three states: An initial state at the beginning of the narrative, the occurrence of an event that leads to a change in the story world and third, the end state after this transformative event (59, 98). These three phases are very similar to the rhetorical structure of nostalgia identified by Stuart Tannock. He describes nostalgia as a "periodizing emotion" and posits three periods that can be identified in nostalgic rhetoric (456–457): Firstly, the "prelapsarian world," namely the Golden Age[2] or glorious past. Second, the "lapse" or "cut," in other words a catastrophe or other events leading to a change. The third period is the "postlapsarian world," i.e. the present that is felt to be lacking or deficient in comparison to the past. Thus, it can be argued that nostalgia follows a narrative structure. This is further supported by Marie-Laure Ryan's broad definition of narrative, which is based on her approach of transmedial narratology. A narrative

---

[1] For a more detailed discussion of the usefulness of employing methods from classical Rhetoric Studies as well as combining them with approaches from British Cultural Studies and Critical Discourse Analysis when studying British political speeches see Göhrmann and Tönnies.
[2] This notion of a lost Golden Age is a further element connecting nostalgia to utopia, more specifically to its origins in Antiquity (Vieira 5).

is thus "the mental construct of a world as well as the textual act of its representation" (6), and it can take any form or medium (6). When one combines Tannock's theoretical approach to nostalgia with the definitions of narrative by Ryan as well as Müller and Grimm, it is not only possible to analyse nostalgic narratives, but also to read nostalgia itself as a narrative, more specifically a narrative about decline. In the following, it will be demonstrated how contemporary political rhetoric constructs such a narrative of nostalgia about Britain's past and present. First, I will focus on the negative representation of the present in Brexit rhetoric before then analysing how it is often contrasted with the construction of the past as a utopian counter-image.

Dystopian Present

Britain's decision to leave the European Union was in part fuelled by a widespread dissatisfaction with the current state of the country on which the leading figureheads of the Leave campaign capitalized in their use of political rhetoric. In fact, the Brexit referendum was represented as an opportunity for the British people to voice their feelings of frustration, loss, and anger by voting Leave (Clarke and Newman 72). As John Clarke and Janet Newman point out, the official campaign slogan "Take Back Control" then offered a "potent promise of redress" for a diverse set of frustrations including but not limited to economic abandonment, cultural dislocation, political disaffection, and nationalist rage (73). In addition to these feelings, the slogan also ties into an underlying "'sense of loss' [...] associated with economic and political, social and cultural and material and affective losses, including the loss of material and psychic 'privilege'" (73). In the following section, I will provide a brief overview of how Brexiters address and instrumentalise the aforementioned feelings of loss, frustration, and discontent with the present in contemporary political rhetoric.

One way in which this is achieved is by acknowledging the personal hardships many people experience due to the recent austerity policies. In her first speech as Prime Minister at a Conservative Party conference, Theresa May for instance, remarked that "it wasn't the wealthy who made the biggest sacrifices after the financial crash, but ordinary, working class [sic] families" ("Keynote"). She then goes on to paint a negative picture of the present in which "household bills rocketed" while people "took a paycut" or even "lost their job" ("Keynote"). Rather than merely providing a neutral description, May employs the artistic proof of *pathos* in her speech as she pointedly makes an appeal to her audience's feelings: "life simply doesn't *seem* fair. It *feels* like your dreams have been sacrificed in the service of others" ("Keynote," emphasis added). This evocation of people's negative emotions about the present state of the country is reinforced by a repetitive structure in which May argues that although "our" society, economy, and democracy "should work for everyone," for many people "it doesn't *feel* like it's working for you" ("Keynote," emphasis added). In doing so, she also depicts the present as a time in which the British population is dealing with rising inequalities. Similar references to high unemployment and people's financial difficulties also frequently occur in political rhetoric of other leading Brexit supporters. One example is Boris Johnson speaking of a "huge difference" between his generation of baby-boomers and that of millennials due to what he describes as a "decline" in "opportunity"; more specifically, "in the scope and power of the younger generation, with their own resources, to buy somewhere to live that they can call their own" ("Party Conference," 2018). Additionally, in an essay announcing his support for the Leave campaign, Michael Gove claimed that "[t]*he euro has created economic misery for Europe's poorest people*" and "*European Union regulation has entrenched mass unemployment*" ("EU Referendum").[3] Such statements are part of a larger nostalgic narrative that portrays Britain's entry to the UK as a cut or lapse leading to the country's economic decline (Johnson, "EU Referendum"; Duncan Smith). In his first major speech during the 2016 referendum campaign, Boris Johnson mentions a lack of innovation[4] or technological advancements as one important factor adding to Britain's struggling economy ("EU Referendum"; see also Gove, "EU Referendum"). Johnson tries to back up the Leave side's nostalgic and emotional argument about the current state of Britain's economy by also making appeals to *logos* as he uses the number of new patents as a measurement for a loss of innovation and cites statistics outlining a decrease in British exports since joining the Common Market ("EU Referendum"). Here, he puts special emphasis on Britain currently not benefiting from existing trade deals as well as being unable to negotiate new ones ("EU Referendum"). Theresa May later picked up on this argument in her Lancaster House speech in which she set out her vision of the country's post-Brexit future as a "Global Britain": "Many in Britain have always felt that the United Kingdom's place in the European Union came at the expense of our global ties, and of a bolder embrace of free trade with the wider world."

May's statement ties in even more closely with another key aspect of the Brexit narrative about Britain's decline, namely the notion that Britain's "influence" on the world stage "will diminish" (Grayling) or that it has lost its former

---

3 For further examples see also Gove ("Risk") and Gove ("Secure").
4 See also Priti Patel's promise of more "freedom [...] to innovate" after Brexit.

role as a leading (colonial) world power. The latter for instance becomes explicit in Conservative MP Owen Paterson's claim that Britain will become "a colony of Europe if we vote to remain, with the Prime Minister reduced to a Roman governor handing down dictats from what Jose Manuel Barroso, former President of the European Commission, described as the 'empire'" (qtd. in Giannangeli) or Boris Johnson equating May's Brexit deal with a "colony status" for which "[p]eople did not vote" ("No One"). There are clear dystopian undertones to these statements which also become apparent in Johnson's widely criticized remarks that the EU "attempt[s] to do [...] by different methods" what "Napoleon, Hitler, various people tried [...] out" in the past (qtd. in Ross). Indeed, there are numerous other instances in which Brexit supporters posit a decline of Britain's "sovereignty" (Gove, "Secure"; Grayling) and "independence" (Johnson, "EU Referendum") as a direct result of the country's membership in the EU. In that regard, Priti Patel even speaks of the "shackles of the EU" (qtd. in Wheeler). This depiction of Britain's relation to the EU is in line with Nora Wenzl's remark upon the use of "words of entrapment" and the prison metaphor frequently employed by Leave supporters (111) or Merle Tönnies' observation of the recurrence of "the image of 'fetters' of which Britain needs to free itself" (151). What further contributes to this dystopian vision of a Britain that has lost "control" (Gove, "Secure"; Johnson, "Party Conference," 2018) of its own affairs is the belief that the EU is responsible for an "erosion of democracy" (Johnson, "EU Referendum"; see also Gove, "Secure"). It is precisely against this dystopian backdrop of a lack of freedom, independence, and democracy in contemporary Britain that the Leave campaign sets its promise of "taking back control."

Another key area where Brexit supporters attest Britain a lack of "control" is within the context of borders and immigration (Raab; Johnson, "EU Referendum"). This is then portrayed as directly contributing to the aforementioned dystopian vision of a decline in people's living conditions[5] and posing a threat to the safety of British citizens due to terrorists and immigrants allegedly being able to enter the country freely (Raab). The dystopian undertones of this rhetoric are reinforced by Raab describing immigrants from the Middle East and Africa as anonymous masses which have "swept across the continent," thereby implicitly

---

5 See for instance Theresa May blaming job losses among the working class on "low-skilled immigration" ("Keynote") or claims that "uncontrolled immigration is putting unsustainable pressures on our public services" like schools or the NHS (Patel qtd. in Doyle; cf. also Johnson "EU Referendum").

evoking the familiar imagery of natural catastrophes like floods.[6] Such a use of nature imagery is also striking in Liam Fox's description of Britain as a "land limited country" and his claim that "[a] constant unchecked flow of migration will inevitably result in more of our open spaces and natural greenery being turned over to housing." Here, it becomes most apparent how issues of borders, control, and national identity are intertwined in Brexit rhetoric's dystopian construction of Britain's present. As nations are typically "imagined as both inherently limited and sovereign" (Anderson 6), open borders and an alleged loss of independence are depicted as restricting Britain's ability to assert its own national identity. In Fox's speech, this is further strengthened by the fact that the statements above also evoke traditional notions of Britishness such as the "green and pleasant land" (Blake 747) or the country's island status. Such threats to British identity or "national wellbeing" (Tönnies 151) are not limited to border issues and Brexit, but are also constructed in other contexts. Strikingly, at the 2020 party conference, Boris Johnson described COVID-19 as a "disease that attacks not only human beings but so many of the greatest things about our country: our pubs, our clubs, our football, our theatre and all the gossipy gregariousness and love of human contact that drives the creativity of our economy." On the whole, the political rhetoric employed by leading Brexit supporters thus paints a dystopian picture of Britain's present which is characterized by declining living conditions, the loss of sovereignty and independence as well as threats to democracy and British national identity.

Utopian Past

In contrast to the negatively evaluated postlapsarian world of the present, the prelapsarian world of the past is often positively and nostalgically evoked in contemporary British political rhetoric. One example is Boris Johnson's 2020 party conference speech, in which he promotes his plan for clean energy production based on wind power by referring to "the history of this country": "It was offshore wind that puffed the sails of Drake and Raleigh and Nelson, and propelled this country to commercial greatness." A similar argument portraying Britain's past as a utopian model for a post-Brexit future was made in 2016 by then-Minister of State for Energy Andrea Leadsom, who, referring to the country's "rich

---

6 Referring to Enoch Powell and Margaret Thatcher as examples, Jonathan Charteris-Black points out that such water imagery has long been used in connection with issues of immigration in British political rhetoric, most notably by the political right (*Metaphors* 147).

history of leadership in energy innovation," specifically mentioned the fact that "[t]he world's first coal-fired power station" and "[t]he world's first commercial nuclear power station" were both located in Britain ("Utility Week"). Such a utopian counter-image to the alleged lack of innovation caused by Britain's EU membership is also constructed in Johnson's 2019 Conservative Party conference speech, where past innovations are listed among other British historical achievements:

> This country has long been a pioneer: we inaugurated the steam age, the atomic age, the age of the genome, we led the way in parliamentary democracy, in female emancipation and when the whole world had succumbed to a different fashion, this country and this party pioneered ideas of free markets and privatisation that spread across the planet. Every one of them was controversial, every one of them was difficult but we have always had the courage to be original, to do things differently.

The reference to free markets similarly evokes a prosperous past in which the country was able to make its own trade deals. In addition, Johnson lists the establishment of Parliament and the emancipation of women in order to construct a continuity between past democratic innovations and Brexit as a means to take back control (cf. also Gove, "Secure"). Thus, he portrays Britain's past as a Golden Age that can serve as inspiration for a better future after Brexit. This statement stands in the rhetorical tradition of Tory politicians, especially Margaret Thatcher,[7] evoking Britain's historical achievements. According to Matthias Göhrmann and Merle Tönnies, "[b]esides inducing pride and in turn uniting her audience under one common identity, [...] [Thatcher's use of this topos] also establishes a stepping stone from which to launch any rhetorical project" (317). As the contemporary examples above show, the listing of British accomplishments fulfils the same function in Brexit rhetoric.[8] This effect is also particularly striking in the speech Jacob Rees-Mogg delivered at a fringe event at the 2017 Conservative party conference:

---

[7] See for instance Thatcher's "The Renewal of Britain" speech where she similarly refers to "scientific and technological innovations" and Britain as "the Mother of Parliaments" as well as positively evoking the British Empire through which the British "have given to the world the English language."
[8] Matthias Göhrmann and Merle Tönnies also remark upon the use of comparable rhetorical strategies in Thatcher's speeches and Brexit rhetoric (316). For further examples demonstrating continuity from Thatcher to the present see Andrea Leadsom's 2016 leadership speech and Theresa May's 2018 party conference speech where Britain is also described as "the mother of all parliaments."

> We need to be reiterating the benefits of Brexit because this is so important in the history of our country. [...] This is Magna Carta, it's the Burgesses coming at Parliament, it's the great reform bill, it's the bill of rights, it's Waterloo, it's Agincourt, it's Crecy. We win all of these things (qtd. in Maidment).

By referring to events connected to the establishment of democracy in Britain, Rees-Mogg similarly evokes the notion of a sovereign Britain where decisions about the country are made by the British people themselves. Thus, he constructs the past as a utopian counter-image to the perceived lack of sovereignty due to the country's membership in the EU in the present. At the same time, he also mentions historical victories, thereby evoking a glorious past in which the British people defended their freedom and triumphed over European states trying to take control of their country.

In fact, Britain's historical wars against other European states are an important element of the nostalgic narrative constructed in Brexit rhetoric (Campanella and Dassú 57–58). This applies especially to Britain's victory in the Second World War. As Edoardo Campanella and Marta Dassú remark, British newspapers such as the *Sun*, the *Daily Telegraph*, the *Daily Express*, and the *Daily Mail* "helped spread war related slogans about Brexit such as 'Independence Day' or 'modern-day Battle of Britain,' or disseminate the spirit of the Blitz and Dunkirk" (61). Such references to the Second World War can not only be found in newspapers, but also in statements by politicians. Whereas such allusions are most prominently used by Nigel Farage,[9] who claimed that "in many ways this referendum on EU membership is our modern-day Battle of Britain," similar parallels were also drawn by representatives of the official Leave campaign and members of May's and Johnson's governments. During the time of the negotiations with the EU, then Brexit secretary David Davis argued that if "our civil service can cope with World War II it can easily cope with this [Brexit]" (qtd. in A. Walker). Similarly, the Conservative politician Jeremy Hunt evokes images of the Second World War as a glorious era of the nation's past when he remarks on the possibility of a no-deal Brexit that "[t]he way Britain reacts is not that we crumble or fold but actually you end up invoking the Dunkirk spirit and we fight back" (qtd. in Gordon). As Malcolm Smith remarks, "Dunkirk" and "the Blitz" "are not just neutral terms, they are totemic. They are the big facts of 1940" (4) which are part of an overarching myth about the Second World War. The cul-

---

[9] Farage's use of nostalgic rhetoric is reinforced by the fact that the campaign bus in which he travelled the country – his so-called "battle bus" – played the theme tune of the popular 1963 war film *The Great Escape*, in which British prisoners of war escaped their German captors (Campanella and Dassú 71).

tural meanings and nostalgic visions of the past associated with this myth of 1940 play such a prominent role in the British historical imaginary that "one need do no more than set these phrases down, without any need to explain them" (Smith 4). Due to these ready-made meanings such phrases can function as effective devices in political rhetoric (Tönnies 149–150). Indeed, the "Dunkirk spirit" referenced by Hunt nostalgically constructs the Second World War as a time in which Britain was standing alone in fighting an external enemy from continental Europe and ultimately able to succeed against all odds (see also Havardi 17–19). This nostalgic construction of a time in which the imagined community was allegedly truly united, offers a utopian counter-image to the widespread divisions in Brexit Britain. Such references to the Second World War are again not limited to Brexit rhetoric but can also be observed within the context of the COVID-19 pandemic. During the first lockdown in March 2020, then Health Secretary Matt Hancock similarly evoked the Blitz spirit to create a sense of unity and foster resilience:

> Our generation has never been tested like this. Our grandparents were, during the Second World War, when our cities were bombed during the Blitz. Despite the pounding every night, the rationing, the loss of life, they pulled together in one gigantic national effort. Today our generation is facing its own test, fighting a very real and new disease. We must fight the disease to protect life ("We Must").

Later that year, Hancock once again nostalgically referred to the Second World War in connection to the pandemic as he made a clear allusion to VE Day by calling the beginning of the vaccination campaign "V-Day" ("It's V-Day").

Another period from Britain's past that plays an important role in connection to Brexit and is often nostalgically evoked in contemporary political rhetoric is the British Empire. In his verdict on the outcome of the referendum, the former leader of the Liberal Democrats Vincent Cable even claimed that Brexit was driven by "nostalgia for a world where passports were blue, faces were white and the map was coloured imperial pink" (qtd. in P. Walker). As Campanella and Dassú remark, politicians from the Leave campaign did not openly speak about restoring the British Empire but rather implicitly or sometimes perhaps even unconsciously evoked this era (67–68). This for instance applies to Theresa May's notion of "Global Britain." In the Lancaster House speech outlining her vision for a post-Brexit future, May remarks that

> we are also a country that has always looked beyond Europe to the wider world. That is why we are one of the most racially diverse countries in Europe, one of the most multicultural members of the European Union, and why – whether we are talking about India, Pakistan, Bangladesh, America, Australia, Canada, New Zealand, countries in Africa or those that are

closer to home in Europe – so many of us have close friends and relatives from across the world.

Here, she only evokes positive memories of the country's colonial history. This becomes even clearer in her claim that "[i]nstinctively, we want to travel to, study in, trade with countries not just in Europe but beyond the borders of our continent." Although May does not openly glorify the Empire and advocate its restoration here, she nonetheless nostalgically portrays it as a Golden Era full of possibilities for Britons in the former colonies and suggests that they had only good intentions like travelling, studying, and trading. In this utopian image of the British Empire, the exploitation of the former colonies and the cruelties of the colonial regime are omitted entirely. A more critical engagement with Britain's past is even rejected outright by Boris Johnson in his 2020 party conference speech. In the context of the renewed discussion about the memorialisation of the British Empire and the colonial slave trade which was sparked by the statue of slave trader Edward Colston being toppled and thrown into Bristol harbour, Johnson remarked that he and his party "are proud of this country's culture and history and traditions" and not "embarrassed to sing old songs about how Britannia rules the waves." In contrast to that, he accused the Labour opposition and implicitly also Black Lives Matter protesters as "literally want[ing] to pull statues down, to re-write the history of our country, to edit our national CV to make it look more politically correct." It can therefore be argued that the utopian image of Britain's past offered by Johnson and other Brexit supporters often provides a totalising view of history that only allows a positive evaluation of the prelapsarian world.

Conclusion: The "Master Narrative of Nostalgia" in Brexit Britain

Due to such claims of absolute truth and omissions of opposing views one can observe a similarity between contemporary nostalgic rhetoric and what Jean-François Lyotard calls "master narratives," i.e. "the big stories, or overarching belief systems, which allow people to organise and interpret their lives and which provide a universal and integrated frame of reference for society. [...] They make absolute, universal and all-embracing claims to knowledge and truth" (Strinati 239). Indeed, one could argue for the existence of what I would call a "master narrative of nostalgia" in contemporary British culture and society. Following Stuart Tannock's rhetorical structure of nostalgia, this master narrative portrays the past positively as a prelapsarian world, positing Britain's loss of its colonies and membership of the EU as a lapse leading to decline and the negatively perceived postlapsarian world of the present. Summarising the findings of this analysis one can ascertain that contemporary British political rhetoric ties into this master narrative of nostalgia: First, Britain's present is depicted in dystopian terms, for instance by claiming a loss of sovereignty or threats to democracy, British national identity, and the country's livelihood. In contrast to that, the past is constructed as a utopian Golden Age characterised by freedom and independence, a sense of national unity, and plentiful opportunities for all Britons. This positive counterimage is constructed through the use of rhetorical devices such as the listing of historical achievements or allusions to the Second World War and the British Empire. As a result, Brexit can thus indeed be seen as indicative of what John Storey calls "utopian nostalgia," with the use of political rhetoric in Brexit Britain perpetuating a dominant master narrative of nostalgia.

Even more so, cultural products also contribute to the construction of the contemporary master narrative of nostalgia and to the close interrelation between political rhetoric, media, and culture in general. In fact, in the last couple of years a great number of historical fictions have featured nostalgic tropes familiar from political rhetoric. Here, the Second World War and the British Empire also seem to play a crucial role as examples like *Dunkirk* (2017), *Darkest Hour* (2017), *Indian Summers* (2015–2016) and *Viceroy's House* (2017) show. In regard to Leavers' feelings of discontent with Britain's present and dystopian visions of further decline in a future without Brexit, the new genre of "BrexLit"[10] also comprises a variety of examples that warrant further analysis. In contemporary British drama, works that engage with nostalgic tropes from the master narrative for instance include Howard Brenton's history play about the Partition of India *Drawing the Line* (2013) or Carol Ann Duffy's and Rufus Norris' verbatim play *My Country: A Work in Progress* (2017) about the Brexit referendum. When analysing how these fictional representations of the country's past and present engage with the master narrative of nostalgia in Brexit Britain, it is of special interest how – apart from reaffirming and supporting nostalgia – they can also function as counter-narratives that subvert and revise it.

---

[10] According to Kristian Shaw, who coined the term, this genre encompasses "fictions that either directly respond or imaginatively allude to Britain's exit from the EU, or engage with the subsequent socio-cultural, economic, racial or cosmopolitical consequences of Britain's withdrawal" (18).

Works Cited

Anderson, Benedict. *Imagined Communities: Reflections on the Origin and Spread of Nationalism.* 2nd. ed. London: Verso, 1991.

Blake, William. "And Did Those Feet." 1804–1810. *The Norton Anthology of Poetry.* Ed. Margaret Ferguson, Mary Jo Salter, and Jon Stallworthy. 5th ed. New York: Norton, 2005. 746–747.

Campanella, Edoardo, and Marta Dassú. *Anglo Nostalgia: The Politics of Emotion in a Fractured West.* London: Hurst & Company, 2019.

Charteris-Black, Jonathan. *Analysing Political Speeches: Rhetoric, Discourse and Metaphor.* London: Palgrave Macmillan, 2014.

Charteris-Black, Jonathan. *Metaphors of Brexit: No Cherries on the Cake?* Cham: Palgrave Macmillan, 2019.

Clarke, John, and Janet Newman. "What's the Subject? Brexit and Politics as Articulation." *Community & Applied Social Psychology* 29.1 (2019): 67–77.

Doyle, Jack. "Schools Are at Breaking Point, Says Employment Minister." *Daily Mail*, 18 Apr. 2016. Web. 24 Jan. 2022. <https://www.dailymail.co.uk/news/article-3544949/Schools-breaking-point-says-employment-minister-Priti-Patel-says-migration-EU-unsustainable-pressure-education-system.html>.

Duncan Smith, Ian. 2016. "Are We in This Together?" *Vote Leave*, 10 May 2016. Web. 4 Aug. 2021. <http://www.voteleavetakecontrol.org/iain_duncan_smith_are_we_in_this_together.html>.

Farage, Nigel. "Farage on Friday: EU Referendum Is Our Modern Day Battle of Britain." *The Express*, 10 July 2015. Web. 10 Aug. 2021. <https://www.express.co.uk/news/politics/590396/Nigel-Farage-European-Union-EU-referendum-Battle-of-Britain>.

Foss, Sonja K. *Rhetorical Criticism: Exploration and Practice.* 4th ed. Long Grove: Waveland, 2009.

Fox, Liam. 2016. "Memories of Green? The Uncontrolled Cost of Immigration." *Vote Leave*, 2 June 2016. Web. 4 Aug. 2021. <http://www.voteleavetakecontrol.org/rt_hon_liam_fox_mp_memories_of_green_the_cost_of_uncontrolled_migration.html>.

Giannangeli, Marco. "EUROPEAN EMPIRE: Powerless Britain to Become Mere COLONY if We Don't Quit Brussels." *The Express*, 24 Apr. 2016. Web. 12 Aug. 2021. <https://www.express.co.uk/news/uk/663846/EU-referendum-David-Cameron-Europe-UK-power-Roman-Governor-Brexit>.

Göhrmann, Matthias, and Merle Tönnies. "British Political Rhetoric from World War II to Brexit: A Cultural Studies Approach." *How to Do Cultural Studies: Ideas, Approaches, Scenarios.* Ed. Jürgen Kramer and Bernd Lenz. Würzburg: Königshausen & Neumann, 2020. 303–339.

Gordon, Rayner. "Jeremy Hunt Warns EU a Bad Brexit Deal Will Stir Britain's 'Dunkirk Spirit.'" *The Telegraph*, 30 Sept. 2018. Web. 12 Aug. 2021. <https://www.telegraph.co.uk/politics/2018/09/30/jeremy-hunt-warns-eu-bad-brexit-deal-will-stir-britains-dunkirk/>.

Gove, Michael. "EU Referendum: Michael Gove's Full Statement on Why He Is Backing Brexit." *The Independent*, 20 Febr. 2016. Web. 11 Aug. 2021. <https://www.independent.co.uk/news/uk/politics/eu-referendum-michael-gove-s-full-statement-on-why-he-is-backing-brexit-a6886221.html>.

Gove, Michael. "The Risk of Remain – Security." *Vote Leave*, 6 June 2016. Web. 5 Aug. 2021. <http://www.voteleavetakecontrol.org/voting_to_stay_in_the_eu_is_the_risky_option.html>.

Gove, Michael. "Secure in Our Values? Why an Unreformed EU Weakens Us in Fighting for Liberal Democracy." *Vote Leave*, 8 June 2016. Web. 4 Aug. 2021. <http://www.voteleavetakecontrol.org/gove_and_raab_eu_membership_makes_us_less_safe.html>.

Grayling, Chris. "We Must Vote Leave to Protect Our Sovereignty and Democracy from Future EU Integration." *Vote Leave*, 31 May 2016. Web. 4 Aug. 2021. <http://www.voteleavetakecontrol.org/chris_grayling_we_must_vote_leave_to_protect_our_sovereignty_and_democracy_from_further_eu_integration.html>.

Hancock, Matt. "We Must All Do Everything in Our Power to Protect Lives." *The Telegraph*, 14 March 2020. Web. 12 Aug. 2021. <https://www.telegraph.co.uk/politics/2020/03/14/must-do-everything-power-protect-lives/>.

Hancock, Matt. "It's V-Day." *Twitter*, 8 Dec. 2020. Web. 11 Aug. 2021. <https://twitter.com/MattHancock/status/1336192262338916352>.

Havardi, Jeremy. *Projecting Britain at War: The National Character in British World War II Films*. Jefferson: McFarland, 2014.

Johnson, Boris. "Boris Johnson's Speech on the EU Referendum: Full Text." *Conservative Home*, 9 May 2016. Web. 11 Aug. 2021. <https://www.conservativehome.com/parliament/2016/05/boris-johnsons-speech-on-the-eu-referendum-full-text.html>.

Johnson, Boris. "Boris Johnson Speech to Conservative Party Conference." *Brexit Central*, 2 Oct. 2018. Web. 4 Aug. 2021. <https://brexitcentral.com/boris-johnson-speech-conservative-party-conference/>.

Johnson, Boris. "No One Is Fooled by this Theatre." *Twitter*, 13 Nov. 2018. Web. 6 Aug. 2021. <https://twitter.com/BorisJohnson/status/1062289237553410048>.

Johnson, Boris. "2019 Conservative Party Conference Speech." *PoliticsHome*, 2 Oct. 2019. Web. 12 Aug. 2021. <https://www.politicshome.com/news/article/read-in-full-boris-johnsons-speech-to-the-2019-conservative-party-conference>.

Johnson, Boris. "2020 Conservative Party Conference Speech." *Conservatives*, 6 Oct. 2020. Web. 12 Aug. 2021. <https://www.conservatives.com/news/2020/boris-johnson--read-the-prime-minister-s-keynote-speech-in-full>.

Leadsom, Andrea. "'I Want to Guide Britain to the Sunlit Uplands' – Full Text of Andrea Leadsom's Leadership Speech." *The Spectator*, 4 July 2016. Web. 4 Aug. 2021. <https://www.spectator.co.uk/article/-i-want-to-guide-britain-to-the-sunlit-uplands---full-text-of-andrea-leadsom-s-leadership-speech>.

Leadsom, Andrea. "Andrea Leadsom Speech to Utility Week Energy Summit." 5 July 2016. Web. 4 Aug. 2021. <https://www.gov.uk/government/speeches/andrea-leadsom-speech-to-utility-week-energy-summit>.

Maidment, Jack. "Jacob Rees-Mogg Compares Brexit to Battles of Agincourt, Waterloo and Trafalgar." *The Telegraph*, 3 Oct. 2017. Web. 12 Aug. 2021. <https://www.telegraph.co.uk/news/2017/10/03/jacob-rees-mogg-compares-brexit-battles-agincourt-waterloo-trafalgar/>.

May, Theresa. "Theresa May's Keynote Speech at Tory Conference in Full." *Independent*, 5 Oct. 2016. Web. 4 Aug. 2021. <https://www.independent.co.uk/news/uk/politics/theresa-may-speech-tory-conference-2016-in-full-transcript-a7346171.html>.

May, Theresa. "Lancaster House Speech 2017." *Time*, 17 Jan. 2017. Web. 12 Aug. 2021. <https://time.com/4636141/theresa-may-brexit-speech-transcript/>.
May, Theresa. "Theresa May's Speech to the 2018 Conservative Party Conference." *PoliticsHome*, 3 Oct. 2018. Web. 4 Aug. 2021. <https://www.politicshome.com/news/article/read-in-full-theresa-mays-speech-to-the-2018-conservative-party-conference>.
Müller, Michael, and Petra Grimm. *Narrative Medienforschung: Einführung in Methodik und Anwendung*. Konstanz: UVK, 2016.
Patel, Priti. "Priti Patel Speech at the Spring Conference of the Association of Licensed Multiple Retailers." *Vote Leave*, 28 Apr. 2016. Web. 4 Aug. 2021. <http://www.voteleavetakecontrol.org/priti_patel_speech_at_the_spring_conference_of_the_association_of_licensed_multiple_retailers.html>.
Raab, Dominic. "The Advantages of Controlled Immigration." *Vote Leave*, 8 June 2016. Web. 4 Aug. 2021. <http://www.voteleavetakecontrol.org/gove_and_raab_eu_membership_makes_us_less_safe.html>.
Ross, Tim. "Boris Johnson. The EU Wants a Superstate, Just As Hitler Did." *The Telegraph*, 15 May 2016. Web. 6 Aug. 2021. <https://www.telegraph.co.uk/news/2016/05/14/boris-johnson-the-eu-wants-a-superstate-just-as-hitler-did/>.
Rowland, Robert C. "The Narrative Perspective." *Rhetorical Criticism: Perspectives in Action*. Ed. Jim A. Kuypers. 2nd ed. Lanham: Rowman & Littlefield, 2016. 125–145.
Ryan, Marie-Laure. "On the Theoretical Foundations of Transmedial Narratology." *Narratology beyond Literary Criticism: Mediality, Disciplinarity*. Ed. Jan Christoph Meister. Berlin: de Gruyter, 2005. 1–23.
Shaw, Kristian. "Brexlit." *Brexit and Literature: Critical and Cultural Responses*. Ed. Robert Eaglestone. London: Routledge, 2018. 15–30.
Smith, Malcolm. *Britain and 1940: History, Myth and Popular Memory*. London: Routledge, 2000.
Storey, John. *Radical Utopianism and Cultural Studies: On Refusing to be Realistic*. Abingdon: Routledge, 2019.
Strinati, Dominic. *An Introduction to Studying Popular Culture*. London: Routledge, 2000.
Tannock, Stuart. "Nostalgia Critique." *Cultural Studies* 9.3 (1995): 453–466.
Thatcher, Margaret. "The Renewal of Britain." *Margaret Thatcher Foundation*, 6 July 1979. Web. 10 Aug. 2021. <https://www.margaretthatcher.org/document/104107>.
Tönnies, Merle. "Constructions of a National Threat: Contextualising the Rhetorical Management of the COVID-19 Pandemic in Britain." *Mentalities and Materialities: Essays in Honour of Jürgen Kamm*. Ed. Philip Jacobi and Anette Pankratz. Würzburg: Königshausen & Neumann, 2021. 149–163.
Vieira, Fátima. "The Concept of Utopia." *The Cambridge Companion to Utopian Literatures*. Ed. Gregory Claeys. Cambridge: Cambridge UP, 2010. 3–27.
Walker, Amy. "Do Mention the War: The Politicians Comparing Brexit to WWII." *The Guardian*, 4 Feb. 2019. Web. 12 Aug. 2021. <https://www.theguardian.com/politics/2019/feb/04/do-mention-the-war-the-politicians-comparing-brexit-to-wwii>.
Walker, Peter. "Vincent Cable Denies Calling Brexit Supporters Racist." *The Guardian*, 12 March 2018. Web. 11 Aug. 2021. <https://www.theguardian.com/politics/2018/mar/12/vince-cable-denies-calling-brexit-supporters-racist>.

Wenzl, Nora. "'The United Kingdom Is a Different State.' Conservative MPs' Appeals to Britishness Before the EU Referendum." *Brexit and Beyond: Nation and Identity*. Ed. Daniela Keller and Ina Habermann. Tübingen: Narr Francke Attempto, 2021. 99–120.

Wheeler, Caroline. "'Our Economy Would Be Better with a Brexit' Employment Minister Blasts EU Membership." *Express*, 10 Apr. 2016. Web. 4 Aug. 2021. <https://www.express.co.uk/news/politics/659613/Britain-UK-economy-Employment-Minister-Priti-Patel-blasts-EU-membership>.

Anette Pankratz
Civil Wars and Republics in Contemporary (Dystopian) Drama

1 Introduction

Once upon a time, there was a kingdom. The people were discontent and fought each other. Eventually, they deposed the monarch and established a republic. To a citizen of the Federal Republic of Germany, this sounds like a fairy tale with a happy ending. In the case of the United Kingdom and its history, however, the case seems to be a bit more complicated. Most of historiography sees the English Civil War and the brief Republic between 1649 and 1660 as failure. More importantly, most British citizens still adhere to a constitutional monarchy as the best form of government.

This article will not delve into comparative cultural studies to find the reasons for these discrepancies. It will rather take the observation as starting point for an analysis of British (dystopian) plays featuring civil wars, riots, and – at least potential – republics, with Rory Mullarkey's *The Wolf from the Door* (2014) as an exemplary and still under-researched case study. The play explicitly and very ironically harks back to the seventeenth century by way of two characters who regularly perform as Roundheads in historical re-enactments of the Civil War, but who are rather squeamish when it comes to staging a revolution. At the same time, Mullarkey invents a rather bloody armed insurrection for the near future of Britain run by a group of Establishment figures.

What can one make of this merging of past, present, and future? How does this tie in with the developments in British drama at large? I will start with a broad overview of British cultural memory and representations of civil wars and (near) revolutions in British drama that represent a divided nation. This will serve as framework for the analysis of dystopian drama in general, zooming in on Mullarkey's play. How revolutionary and subversive is its post-Brechtian form? What are the ideological ramifications of its strategies of representation? What can one infer from this about British culture in general?

2 History, Historiography, and Drama

The English Civil War and the Republic are part of two narratives with different political connotations. One strand, associated with Marxist and neo-Marxist

https://doi.org/10.1515/9783110758252-011

scholarship, highlights the unfulfilled utopian potential of the events. Historian Christopher Hill, for example, emphasises the significance of the "revolt within the revolution" as a moment in history, when "[l]iterally anything seemed possible" (14). Democratic and proto-Communist groups like the Levellers, Diggers, or Ranters "might have established communal property, a far wider democracy in political and legal institutions, might have disestablished the state church and rejected the protestant ethic" with its early capitalist "ideology of the men of property" (15).

The more dominant narrative in British culture, on the other hand, sees the time between 1642 and 1660 as relative failure. In this version, the British success story starts with the "Glorious" Revolution of 1688/1689, which managed a change of monarchs without endangering the hegemony of the propertied elites. This is a central part of the Whig Interpretation of history, made famous by Thomas Babington Macaulay who claimed that "the history of our country during the last hundred and sixty years is eminently the history of physical, of moral, and of intellectual improvement" (3). A felicitous collaboration of the ruling classes, a series of Protestant monarchs, and God (aka "Him") guaranteed that Britain was spared the chaos of a genuine revolution and could thus secure stability and prosperity:

> It is because we had a preserving revolution in the seventeenth century that we have not had a destroying revolution in the nineteenth [...] our gratitude is due under Him who raises and pulls down nations at His pleasure, to the Long Parliament, to the Convention and to William of Orange. (448)

Theatre and drama tend to appropriate the Marxist narrative and thereby provide a counter-narrative to the dominant Whig Interpretation of history. Directly influenced by Hill's seminal study *The World Turned Upside Down* (1975), Caryl Churchill's *Light Shining in Buckinghamshire* (1976) presented the utopian ideas of the Diggers, Levellers, and Ranters. In her introduction to the play, Churchill echoes Hill in describing the "revolt in the revolution" as a time "when the king had been defeated and anything seemed possible, and the play shows the amazed excitement of people taking hold of their own lives" (183; cf. Roberts and Stafford-Clark 24).

The play's revolutionary impetus also manifests itself in its form: multiple casting, a fragmented plot, and Brechtian alienation effects. Likewise, on the meta-level, the co-operative and collective methods of theatre groups like Joint Stock, the workshops and discussions between director, actors, and author (and later between performers and the audience), installed the theatre as utopian (performative) space in the sense of Jill Dolan (Adiseshiah).

3 Divided Nations

The English Civil War returns on stage in the 2010s, albeit in slightly different forms. *Light Shining in Buckinghamshire* has a revival at the National Theatre in 2015 in a luscious production, foregoing most of its Brechtian elements. Howard Brenton's *55 Days* (2012) personalises the Civil War (without completely shirking its complexities) as a conflict between Charles I and Oliver Cromwell – the childish king who stubbornly insists on his divine right and the religiously inspired, but also ruthlessly pragmatic politician. At the end, the latent utopia – "fighting for a new Jerusalem" as the nameless Third Trooper puts it (7) – is foregone in favour of political power and a stable system. Indirectly, the play seems to endorse the status quo of 2012. In hindsight, Cromwell's throwaway remark "[i]magine a time when no one takes monarchy seriously, and we all just get on with living" (33) and his offer to Charles to become a constitutional monarch (87–88) describe the best compromise formation to save the King's life and empower Parliament. It was not taken in 1648, but is available now.

The notion of England as new Jerusalem also looms large in Jez Butterworth's *Jerusalem* (2009), considered by most critics as the paradigmatic state-of-the-nation play of the early 2000s (Sierz 141; Edwards 281). The protagonist Johnny "Rooster" Byron calls for a "Flintock Rebellion" (Butterworth 53) and envisions that "we will storm Flintock Village and burn every house, shop and farm. We will behead the Mayor. Imprison the Rotary Club. Pillage the pubs! Rob the tombola! And whip into a whirlwind a roughhead army of unwashed, unstable, unhinged, friendless, penniless, baffled berserkers" (52). Nothing comes of it, and at the end of the play, Rooster has to give in to the authorities of the Kennet and Avon council. Nevertheless, *Jerusalem* as a whole celebrates Rooster's anarchic energy and his carnivalesque resistance against the neoliberal present.

Setting "them" against "us" – elites against the "unwashed, unstable, unhinged" – recurs in quite a few contemporary plays based on the notion of the UK as "divided nation" along the lines of class, gender, ethnicity, religion, and region – to mention the most pertinent factors (Sierz 127). In the last years, these divisions have crystallised in the discourses about Brexit, but the rift goes further back, to Thatcherism, devolution, the debates about institutional racism, and austerity (Rebellato, "Nation and Negation" 16).

In drama, the dis-united society often manifests itself in riotous behaviour – across the classes –, from the London Riots of 2011 in Alecky Blythe's verbatim *Little Revolution* (2014) to the fictionalised depiction of the Bullingdon Club in Laura Wade's *Posh* (2010/2012). As in *Jerusalem*, the riots threaten to destabilise the status quo and to overthrow the system, but all of these plays operate with a more or less realist frame: the country may be divided, but the riots do not spread all over the nation and develop a revolutionary dynamic. Not so in works like Mike Bartlett's *Charles III* (2014), Mullarkey's *The Wolf from the Door*, or Alice Birch's *Revolt. She Said. Revolt Again* (2014/2016). These dystopian dramas confront the audience with more or less explicit, more or less successful scenarios of civil war.

4 Dystopian Plays and Civil Wars

Dystopias, "fictional [...] places set apart from the real world and openly marked out as 'bad'" (Sierz and Tönnies 20), extrapolate contemporary trends and project them into the future or into a parallel world in order to warn of current sociopolitical developments, thereby providing a counter-position to hegemonic structures (Baccolini and Moylan 7). Dystopias have dominated twentieth- and 21st-century literature and mass media. Theatre and drama discovered the sub-genre rather late, but since the 1990s, dystopias have been flourishing on the stage as well (Voigts and Tönnies; Sierz and Tönnies).

Quite a few dystopian plays employ scenarios based on or culminating in civil wars. One of the best and most often quoted examples is Churchill's *Far Away* (2000). In the last part, Joan envisions a Hobbesian war of everyone against everyone: nationalities, professions, animals, and elemental forces: "the weather here's on the side of the Japanese" (37; cf. Sierz and Tönnies 22). Equally famous, Sarah Kane's *Blasted* (1995) depicts a conflict reminiscent of the war in the former Yugoslavia waged in a hotel room in Leeds. While the dystopian scenarios by Churchill and Kane appropriate widely mediatised images of actual civil wars, the dystopian settings in plays of the 2010s appear less specific. At the same time, they contain more explicit references to political resistance and agency (Reid 82). Birch's *Revolt. She Said. Revolt Again* reiterates the imperative to "revolutionize" language, work, the world, and the body and ends with a bucket list for revolutionary change:

- We're going to dismantle the monetary system [...]
- And overthrow the government [...]
- All jobs will be destroyed
- And all couples broken

– And we take over the airwaves, the televisions, the Internet, etcetera.
.¹
– And we'll eradicate all men. (74)

At the end of the play, however, the future gives way to the past tense and an admission of failure: "It failed. The whole world failed at it. [...] Who knew that life could be so awful" (75). The Beckettian overtones of the title, however, imply that the tone of resignation does not mean the end of attempts at revolution: "Try again. Fail again. Fail better."

Bartlett's *King Charles III* focuses on the upper echelons of political power with the real Windsors as protagonists. By refusing to sign a Bill restricting the freedom of the press, the new monarch Charles III provokes a civil war. The King resembles Charles I in many respects. He evokes the myth of the sacred body politic and does not want to accept a parliamentary system: "there is / A wise and ancient bond between the Crown / And population of this pleasant isle. / It's only in the last five hundred years / That politicians and democracy / Have led the way in policy and meant / The people vote for who they want to lead" (64–65). Like Charles I in 1642, Charles III also enters the House of Commons and dissolves Parliament. But – in contrast to the seventeenth century – no revolution ensues. The monarchy is saved by the King's abdication and handing over the crown to his son William. Again, like *55 Days*, *King Charles III* implicitly promotes the status quo of the present as compromise formation, singing the praises of constitutional monarchy with allusions to Walter Bagehot (108) and admiration for Elizabeth II: "Maybe she was what held it together" (79).

5 *The Wolf from the Door*

Mullarkey's *The Wolf from the Door* uses a similar constellation – a young charismatic King ends a civil war and brings peace and prosperity –, albeit in a less realistic mode and with much more irony. Trish Reid and Dan Rebellato highlight its revolutionary impetus and the way it pushes "at the edges of realism," offering scenes full of "absurd, comic acts of violence" (Rebellato, "Apocalyptic Tone" para. 30). The absurd and comic premise of the play – the urban elites and the bland middle classes stage a successful uprising – seems to appropriate and reverse Rooster's vision of the "Flintock Rebellion" in *Jerusalem*. The Morris dancers who in Butterworth's play represent dead and commodified tradition,

---

1 Not a typo, but stipulating a pause, "*whether this is a single beat or ten minutes depends on what feels right*" (19).

here become a vital part of the revolution: "A hollering mass / Of white-clothed figures" (39). Instead of a lord of misrule from the marginalized so-called underclass and his group of "baffled berserkers," the civil war is carefully planned by Lady Catherine Dean and her friends in high places. Despite these differences, *Jerusalem* and *The Wolf from the Door* share the same critical stance against "the political, commercial and social institutions" of the UK, as Catherine puts it (31).

In contrast to the dystopias by Birch and Churchill, *The Wolf from the Door* offers a happy ending of sorts. After the successful insurrection, Leo Lionheart becomes the new King. With his Black mother and an unknown, probably non-British father, and with his social status as homeless orphan, Leo represents both a multicultural Britain and the voice of the abject. Under his rule, "everything is brilliant":

> Courtier One: Crime is down.
> Courtier Two: Happiness is up.
> Courtier One: The wealth is evenly distributed.
> Courtier Two: And public services are running well. (47)

Admittedly, the ending has to be taken with more than just a pinch of salt and this also applies to the staging of the great revolution. But, as I will try to show, the play's ideological underpinnings point towards a double-edged use of the past – vying between nostalgia, retrotopia, and utopia – that *The Wolf from the Door* shares with *Jerusalem* and quite a few other plays and which also shapes discourses in contemporary Britain.

5.1 Interrelations between Past, Present, and Future

Both utopian and dystopian texts tend to hark back to the past in order to envision a better future and to point out the faultlines of the present. In William Morris's *News from Nowhere* (1890), time traveller William Guest finds himself in a post-revolutionary, socialist future where people have returned to the clothes and the economy of the Middle Ages. Winston Smith seeks consolation in snippets of old songs and outdated memorabilia in George Orwell's *Nineteen Eighty-Four* (1949). Even if his private archaeology of the past does not help to liberate Oceania, it brings temporary respite and bolsters Winston's will to resistance.

As John Storey has pointed out, utopian desire – "the depiction or enactment of something that currently does not exist, or does not yet exist in a fully developed form, in order to incite the desire for it in the here and now"

("Radical Unfolding" 107) – can look back to the past in order to unearth fruitful options for changing the status quo (*Utopianism* 55). Storey situates important impulses for this endeavour in the seventeenth century and the English Revolution. He singles out the Diggers, whose ideal of the common ownership of land, common labour, and the radical equality of people imitated early Christian communities and thereby threatened to undermine social and political hierarchies. The utopian desires of the Diggers were in turn appropriated by the counter-cultures of the 1960s and 1970s.

The Wolf from the Door also merges past, present, and future, although its utopian desire appears more oblique. By way of his new chosen name, Leo Lionheart refers back to the almost mythical British monarch of Robin Hood and *Ivanhoe* fame. In ascending the throne, he realises the revolutionary ideal of a world upside down, bringing the marginalized Other centre stage and into a position of power. He has been brought there, however, by Catherine and her upper-class friends, who chose him as King because "[y]ou are of uncertain parentage, no fixed abode and no employment. You have no education, no qualifications, no personal ties and no possessions. You exist outside society" (35). In contrast to Leo, the rest of the revolutionaries are very much part of society and of British tradition. Catherine herself was born into the aristocracy; her father was "[o]ne of the Marquesses. So I'm a 'Lady'" (7), and she has inherited quite a bit of "old money" (5). Her close circle of conspirators consists of the Bishop of Bath and Wells, society lady Emily-Jane Sorenson, and choir master Pemberton, who indulge in flower-arrangement, hymn-singing, and armed insurrection (13).

While resistance in classical dystopias is usually directed against a totalitarian regime, the revolutionaries in *The Wolf from the Door* have a much more amorphous adversary: the neoliberal system (Reid; Rebellato, "Apocalyptic Tone"). Catherine's first target is the rather randomly picked assistant manager of a supermarket, because:

> The problem, Derek, is that I feel utterly powerless. [...] and I feel like your supermarket is one of the key causes of this feeling. So I am here seeking compensation. [...] I want your life. (10)

At Catherine's behest, Leo decapitates Derek with a sword.

Neoliberalism and consumerism appear to be the root of all evil. They stand for a lack of agency, a cancerous ideology that implements the rules of the market in all realms of life and thus exhausts people through the constant need to compete. The loquacious Minicab Driver tells his passengers Catherine and Leo about his daughter who suffers from depression: "My daughter's just done her exams and she just lies there in her bed, eyes on the ceiling, won't leave her bed, won't change her posture, just looks at the ceiling" (23). A bit later, he rambles on about "the market, the market, the way, the way, the way, the way, the market, the way" (23).

The crux of the matter, as the Bishop of Bath and Wells points out, is that there is no viable alternative. More precisely: as neoliberalism permeates all realms of life and thus has to be taken as reality, people are unable to imagine any alternatives: "We have reached an end point. A point where a shameful, cruel and hypocritical system perpetuates itself because it instils in us, from birth, a belief that there is no alternative" (36).

In the world of the play, "capitalist realism," a concept coined by Mark Fisher, has become a truism. Catherine's criticism is rather vague and consists of snappy soundbites. She and her friends supposedly fight against "a system which preaches moderation and fairness and equality but in fact merely breeds division and slavery and poverty" (31). Ironically, they are not directly and personally concerned by "slavery and poverty," and the division is shown to derive more from their machinations than from the activities of the likes of Derek. Moreover, Catherine and her friends do not intend to take over power (and thereby also responsibility). On the contrary, the Bishop plans to commit suicide, and Catherine wants to leave the country: "I'm a Marquess's daughter. I'm like the Bishop. One of the old lot. There's no place for me now" (43).

The English Civil War, or rather its position in British cultural memory, is equally made fun of by way of two re-enactors. As with the insurrectionists, the play here gets a lot of comic mileage out of the discrepancy between middle-class respectability and insurrectionary furor. The two Roundheads Jules and Judith are friendly, chatty, and obtrusive. They would prefer not to get involved in Catherine's revolution, because:

> Jules: It just doesn't really sound like our type of thing. [...]
> Catherine: Armed insurrection isn't your type of thing?
> Jules: Not really, no.
> Leo: So what is your type of thing?
> Judith: Erm. English Civil War? (29)

In the end, they can be cajoled into joining ("if you do not, you will die" 32). But, all in all, the historical revolution seems to have regressed into a performance devoid of meaning.

The representation of the present revolution in *The Wolf from the Door* follows similar patterns of comic incongruity. Violent attacks on the symbols of the Establishment are connected with harmless upper-middle class pastimes:

> *Buckingham Palace is raided by an over-seventies golf team. [...]*
> *Westminster Abbey gets napalmed by a ceilidh group.*
> *A water-polo team shoot their rocket launchers at Ten Downing Street. [...]*
> *The BBC is bulldozed by South London Cossack Dance Society. [...]*
> *Some theatres get firebombed by a lawn bowls association.*
> *A poker syndicate launch cruise missiles at an array of Oxbridge colleges.*
> *The Houses of Parliament explode.* (42–43)

It is unclear who is fighting whom. The synecdochal loci of power are associated with the traditional elites, "politicians," and "our bosses." At the same time, the revolutionaries also attack groups which are more or less powerless: "hipsters," "football players," or "film directors" (41). The anonymous first-person narrative in scene thirteen points out that "[t]he sick, subdued and silent mass of England / Rises up" (41), mentioning "night club bouncers," "Polish builders," and "unpaid interns" (41). This ties in with the allusions to the London Riots and the Occupy Movement: "When all those shops got smashed / Or when those people camped outside St Paul's" (41). The "liberal pundit" interprets these public protests as forerunners of the revolution: "he's written several books about this subject / And says he saw it coming a mile off or something" (41). But even in this passage, it is mostly members of the "Morris-dancing association" (39), students of intermediate German (41), women from a hot yoga group, and "ladies who lunch" (41) who engage in the uprising.

5.2 Ukania

What can one make of this? Where is the problem? Where the solution? Where the dystopian dangers and where the utopian desire? *The Wolf from the Door* maintains an ironic stance and does not commit itself to a clear political position. The middle classes are made fun of, and so is their attack on both the neoliberal system and old traditions. If one focuses on the trajectory of the plot – from revolution to a glorious restoration –, one can detect traces of the Whig Interpretation of history: not the unwashed masses, but the discerning middle class led by the old elites is able to bring about a positive, "preserving" revolution with the monarchy and social hierarchies still intact. The state of the nation at the end of the play comes close to what Tom Nairn calls *Ukania*,[2] a classbound, pre-modern country, devoted to its monarchy and the old regime (97). Ac-

---

[2] A reference to *Kakanien*, the scatological term for the Austrian imperial monarchy in Robert Musil's novel *Der Mann ohne Eigenschaften* (1930).

cording to Nairn, Britain's political system is basically a remnant of "early-modern times" (97), run by an "*essentially* hereditary élite" (97), "aristocratic and family-based [...,] uniting landownership with large scale commerce and managing 'its' society as a cooptive estate" (152), with the monarchy at the apex of the pyramid, both mythical and popular. Nairn claims that Ukania proper ended in 1979 with Margaret Thatcher's neoliberal government, which shook up the quasi-natural predominance of the old elites (137). The fascination with the monarchy and the popularity of all things royal remained, though, and kept the system stable.

From this perspective, Catherine's attempts at a regime change and Leo Lionheart's new Kingdom can be read as a reflex of Ukania. The monarchy is as popular as ever, although or even because Leo does not seem qualified for the job at all – just like the Windsors. In a satirical take on the myth of Royal extraordinary ordinariness (Nairn 45), Leo is an Everyman figure imbued with messianic traits. Exaggerating the mediatised images and popular notions of the members of the Royal family, Leo seems above banal human needs: "I don't need to sleep or eat or drink. [...] I only have one set of clothes because those ones never really smell because I never really sweat. I don't need to wash" (Mullarkey 7). The Roundheads later corroborate his special aura when they observe that he looks like Charles I, Thomas Fairfax, "[o]r one of those other English Civil War guys" (26) and assume that there is "royal blood" in him (27). As the Bishop of Bath and Wells reminds Leo, he is the only one to "offer up any kind of true alternative" (36), because he is an outsider. One could read this as empowerment and agency for the abject, but Leo becoming monarch – and not President, revolutionary leader, or part of an anarchist collective – rather turns him into a token.

Where plays of the 1980s and early 1990s such as Brenton's *Greenland* (1987) or Edward Bond's *War Plays* (1985) would commit themselves and present a socialist utopia on stage, *The Wolf from the Door* ends with a new monarch as doubtful, clearly non-realistic alternative (marked by the stage directions as "*bizarre*" 45–46). Ukania and neoliberalism are paradoxically intertwined with a civil war that is waged by the elites for the elites against the elites.

This correlates with developments in utopian and dystopian writing: "the post-Fordist, fluid-modern world of freely choosing individuals does not worry about the sinister Big Brother" (Bauman, *Liquid Modernity* 61; cf. Fisher 1). Global capitalism has become hegemonic and without alternative, "nothing new can ever happen" (Fisher 4). In this context, the turn towards the past attains nostalgic overtones and produces what Zygmunt Bauman calls "retrotopias," cherishing close-knit communities, heritage, and tradition and producing "visions located in the lost/stolen/abandoned but undead past, instead of being tied to the not-yet-unborn and so inexistent future, as was their twice-removed forebear" (Bauman, *Retrotopia* 5).

This ties in with developments in British culture. Both James Bond and Boris Johnson evoke nostalgia by marking the Second World War as Britain's "finest hour." In the context of Brexit, the "Glorious" Revolution and thereby also a return to Ukania, play an equally important role. In 2017, Jacob Rees-Mogg claims: "Leaving the European Union is a great liberation for the United Kingdom, as worthy for celebration as victory at Waterloo or the Glorious Revolution" (qtd. in Flood) – a theme reiterated by Johnson in his 2020 Greenwich speech and by political commentators right before the final Brexit exit in 2021. Again, Brexit serves as catalyst and not as cause for these looks back in admiration. Moreover, not only Conservatives tend to celebrate the past. As Alistair Bonnett and Nick Cohen, amongst others, have pointed out, nostalgia permeates British politics from both the Left and the Right.

It also permeates British plays. *Jerusalem* celebrates a traditional and popular "Deep England," and culminates in Rooster inscribing himself into the community of mythical ancestral giants (Edwards 286). This can be read as liberating and Dionysian, but, as Gemma Edwards points out, "Deep England" also essentialises and excludes by its emphasis on "blood and soil" (286). This correlates with Englishness being constructed as representative of the UK as a whole. Set in Wiltshire and using St George's Cross as significant element of the stage set, *Jerusalem* highlights its Englishness, and so does *The Wolf from the Door*. The characters do not leave the South West of the country, and London and Oxford represent the nation at large. Positioning England as equivalent of the whole nation is foregrounded even more in Mullarkey's *St George and the Dragon* (2017), featuring the English patron saint as its hero.

Catherine's upbringing into a position of entitlement and her predilection for violence connect *The Wolf from the Door* with the Riot Club in Wade's *Posh* and its ambivalent depiction of Ukanian youths.[3] Goaded on by a vision of Lord Riot to "[t]ear it down and build something better" (115), the "boys," all members of an Oxford dining club, wreck the place where they are having dinner and beat up the landlord. Although the characters all come from the upper class with huge amounts of social, cultural, and economic capital, their diction and their values – drinking, partying, smashing things up – resemble those of Rooster and his gang. And the same destructive, utopian energy can be found in *The*

---

3 Many thanks to Vicky Angelaki for pointing out the parallels between *Posh* and *The Wolf from the Door*. And while I am in the footnote for appreciations, I would also like to express my gratitude to the participants and the workshop organisers for the productive and inspiring papers and discussions.

Wolf from the Door. Catherine revels in decapitation, shooting, and maiming. In order to convince the two Roundheads not to shirk the insurrection, she sings the praises of "the beautiful violence which brings change […,] a better future for our children" (32). Is this part of the utopian desire described by Storey? Or the equivalent of Slavoj Žižek's "divine violence," perpetrated to mark injustice and inequality (169)?

5.3 Post-Brechtian Comedy

Of course not. *The Wolf from the Door* is an "anarchic satire" (Reid 73) that cannot be taken at face value. Its strategy of the dystopian defamiliarisation (Booker 19) of civil wars and revolutions is comparable to Monty Python's re-enactment of Pearl Harbour by the Batley Townswomen's Guild. Harmless, boring people turn into raging maniacs. Rebellato, Reid, and Elisabeth Angel-Perez point out the efficacy of the comic alienation effects. Where capitalist realism allows no alternatives and claims to base its ideologies on truth and an objective view of reality, dystopian plays confront the audience with what is not. They always also bear traces of utopia – whether directly by imagining alternatives to the "bad place" or indirectly by way of default – and hence potentially feed into the imagining of alternatives to the present, at least offering a "horizon of hope" (Baccolini and Moylan 6).

The Wolf from the Door and quite a few other plays frame their dystopias as comic or satirical. The alienation effects rely on comic incongruities and carnivalesque reversals: the staid middle-classes turn revolutionaries; the abject move to the apex of the sociopolitical pyramid. Leo's frequent tears become the source of laughter and black humour (11; 48). But the comedy also destabilises the criticism. Neither the old elites with their facile soundbites nor the representatives or the victims of neoliberalism can be taken seriously. Derek does not live long enough to develop a personality of his own; the Minicab Driver's monologue encapsulates the drabness of his life.

How far does the subversive potential of comedy and non-realism go?[4] Liz Tomlin concedes that irony can help to undermine the interpellations of the mainstream, "to reveal the naturalized as ideological and to induce self-reflection and self-critique on the part of the spectator" (80). Likewise, Angel-Perez

---

4 In the following passage, I will be mixing the concepts of comedy, satire, irony, and humour for the sake of the argument. I am aware of the differences, but see incongruity and frame-breaking as their essential common denominator.

highlights the political function of the "neo-satiricist ethos which we may call "horrendhilarious" (5). By going to absurd extremes, verbal wit fuses comedy with tragedy, and this "apocalyptic laughter" (Kristeva qtd. in Angel-Perez 8) makes the spectators see and realise deficits in current culture.

Comedy, however, also creates in-groups and out-groups, those who "get" the joke (and are able to laugh about it), and those who do not (Tomlin 86; Berlant and Ngai 235). With its many layers of comic incongruity, *The Wolf from the Door* relies on a well-informed audience "in the know," identifiable as members of the intellectual middle-class who frequent the Royal Court Theatre (Tomlin 94). Moreover, comedy and humour may always be political, but they are not necessarily on the side of the non-normative or subversive. Racist, sexist, homophobic, or antisemitic jokes laugh down at marginalized groups and clearly indicate a position of power (Berlant and Ngai 247; Tomlin 94–95).

In its "hypertrophy of violent imaginative representation" (Rebellato, "Apocalyptic Tone" para. 29), *The Wolf from the Door* tests the limits of comedy. This strategy fits in with the current "age of awkwardness" and cringe characterised by an array of different, often contradictory values and norms (Kotsko 7–8). But, to argue with Fisher and Bauman, while there is no longer a clear-cut set of norms for our daily lives, sexualities, forms of living together, or the organisation of work, neoliberalism does promote the norm of success (or of at least making an effort). Criticism and resistance are an integral part of the system; irony and cynicism not only make it possible to accept and endure the internalised pressure to succeed, but also promote flexibility as the "super-norm" (Böhn 55; cf. Fisher 5; 12–15; Berlant and Ngai 237).

The ironic stance of *The Wolf from the Door* as well as its complexity feed into this detached attitude. The alienation effects fuel an ambivalent reaction in the spectators, as described by Tomlin: "in a state of detached amusement, there is the risk that the precarious spectator-subject might be frozen by the dystopia they are confronted with and their own incapacity to know what they can possibly do to prevent it" (96). We are amused in an intellectually stimulating way, but do not feel the need to get engaged.

6 Conclusion

The characters in *The Wolf from the Door* are stuck between the Scylla of neoliberalism and the Charybdis of Ukania. Despite their claims to the contrary, Catherine and her friends are the only ones with agency – thanks to their old money which also makes the neoliberal world go round. The insurrection and Leo's monarchy replay the "Glorious" Revolution. At the same time, going back – either to the Civil War or the times of the old elites – seems ridiculous. *The Wolf from the Door* both evokes and negates revolution. It does not present alternatives, but a shrug and a laugh. And then we carry on.

This correlates with quite a few other recent plays. Despite their differences in dramatic form, the sociopolitical situation of characters, and their themes, *Jerusalem*, *Posh*, and *Charles III* rely on traditional concepts of national identity and hegemony. Moreover, to come back to the observation from the beginning of the article, those seem to be unthinkable without the monarchy. Notions of an early-modern community led by a benevolent authority figure, preferably a monarch, can also be found in Carol Ann Duffy and Rufus Norris's Brexit play *My Country: A Work in Progress* (2017). People contradict and oppose each other, but they can agree at least on one facet of British life: "At the end of the day we've got Queen, Queen and country" (19). Most of them distrust the politicians and the elites; some jokingly wish for a "benign dictator" (56), but at the end of the play, Britannia admonishes: "we should seek and search and strive for good leadership" (58).

On the meta-level this also begs questions about the efficacy of theatre as institution – how realistic are claims about its utopian or democratic potential? Recent disclosures about dictatorial (or monarchical) directors, sexism, racism, and a general misuse of power, as well as the pressure on artists to bear often inhuman working conditions and sub-standard pay, imply that these assertions are as fictitious as the revolutions and civil wars shown on stage (Harvie 163–164). The divided nation, however, looks as real as ever.

Works Cited

Adiseshiah, Siân. "Utopian Space in Caryl Churchill's History Plays: *Light Shining in Buckinghamshire* and *Vinegar Tom*." *Utopian Studies* 16.1 (2005): 3–26.

Angel-Perez, Elisabeth. "'In the Heart of Each Joke Hides a Little Holocaust' (George Tabori): Horrendhilarious Wit on the British Contemporary Stage." *Miranda* 19 (2019): 1–11. Web. 8 March 2022. <https://doi.org/10.4000/miranda.19898>.

Baccolini, Raffaela, and Tom Moylan. "Introduction: Dystopia and Histories." *Dark Horizons: Science Fiction and the Utopian Imagination*. Ed. Raffaela Baccolini and Tom Moylan. New York: Routledge, 2003. 1–12.

Bartlett, Mike. *King Charles III*. London: Nick Hern Books, 2014.

Bauman, Zygmunt. *Liquid Modernity*. Cambridge: Polity Press, 2012.

Bauman, Zygmunt. *Retrotopia*. Cambridge: Polity Press, 2017.

Berlant, Lauren, and Sianne Ngai. "Comedy Has Issues." *Critical Inquiry* 43 (2017): 233–249.

Birch, Alice. *Revolt. She Said. Revolt Again*. London: Oberon Modern Plays, 2014.

Böhn, Andreas. "Subversions of Gender: Identities through Laughter and the Comic?" *Gender and Laughter: Comic Affirmation and Subversion in Traditional and Modern Media*. Ed.

Gaby Pailer, Andreas Böhn, Stefan Horlacher, and Ulrich Scheck. Amsterdam: Rodopi, 2009. 49–64.
Bonnett, Alistair. *Left in the Past: Radicalism and the Politics of Nostalgia*. New York: Bloomsbury Academic, 2010.
Booker, M. Keith. *The Dystopian Impulse in Modern Literature: Fiction as Social Criticism*. Westport: Greenwood, 1994.
Brenton, Howard. *55 Days*. London: Nick Hern Books, 2012.
Butterworth, Jez. *Jerusalem*. London: Nick Hern Books, 2009.
Churchill, Caryl. Introduction. *Light Shining in Buckinghamshire. Plays One*. London: Methuen, 1985. 183.
Churchill, Caryl. *Far Away*. London: Nick Hern Books, 2000.
Cohen, Nick. "Our Politics of Nostalgia is a Sure Sign of Present-day Decay." *The Guardian*, 26 June 2021. Web. 13 Oct. 2021. <https://www.theguardian.com/commentisfree/2021/jun/26/our-politics-of-nostalgia-is-a-sure-sign-of-present-day-decay>.
Duffy, Carol Ann, and Rufus Norris. *My Country: A Work in Progress*. London: Faber, 2017.
Edwards, Gemma. "This is England 2021: Staging England and Englishness in Contemporary Theatre." *Journal of Contemporary Drama in English* 9.2 (2021): 281–303.
Fisher, Mark. *Capitalist Realism: Is There No Alternative?* Winchester: Zero Books, 2009.
Flood, Alison. "'A Vote for Freedom': Jacob Rees-Mogg Joins Lionel Shriver and Matt Haig in Brexit Anthology." *The Guardian*, 20 Sept. 2017. Web. 10 March 2021. <htpps://www.theguardian.com/books/2017/sep/20/jacob-rees-mogg-lionel-shriver-matt-haig-brexit-anthology-goodbye-Europe>.
Harvie, Jen. "The Power of Abuse." *The Routledge Companion to Theatre and Politics*. Ed Peter Eckersall and Helena Grehan. London: Routledge, 2019. 163–167.
Hill, Christopher. *The World Turned Upside Down: Radical Ideas During the English Revolution*. Harmondsworth: Penguin, 1975.
Kotsko, Adam. *Awkwardness*. Winchester: Zero Books, 2010.
Macaulay, Thomas Babington. *The History of England from the Accession of James the Second*. Vol. II. London: Longman, Brown, Green, and Longmans, 1849.
Mullarkey, Rory. *The Wolf from the Door*. London: Bloomsbury Methuen Drama, 2014.
Nairn, Tom. *The Enchanted Glass*, London: Hutchinson, 1988.
Rebellato, Dan. "Of an Apocalyptic Tone Recently Adopted in Theatre: British Drama, Violence and Writing." *Silages Critiques* 22.1. (2017). Web. 8 March 2022. <https://doi.org/10.4000/sillagescritiques.4798>.
Rebellato, Dan. "Nation and Negation (Terrible Rage)." *Journal of Contemporary Drama in English* 6.1 (2018): 15–39.
Reid, Trish. "The Dystopian Near-Future in Contemporary British Drama." *Journal of Contemporary Drama in English* 7.1 (2019): 72–88.
Roberts, Philip, and Max Stafford-Clark. *Taking Stock: The Theatre of Max Stafford-Clark*. London: Nick Hern Books, 2007.
Sierz, Aleks. *Rewriting the Nation: British Theatre Today*. London: Methuen, 2011.
Sierz, Aleks, and Merle Tönnies. "'Who's Going to Mobilise Darkness and Silence?': The Construction of Dystopian Spaces in Contemporary British Drama." *Journal of Contemporary Drama in English* 9.2 (2021): 20–42.
Storey, John. *Radical Utopianism and Cultural Studies*. London: Routledge, 2019.

Storey, John. "A Radical Unfolding: Utopianism against Complicity." *Complicity and the Politics of Representation*. Ed. Cornelia Wächter and Robert Wirth. London and New York: Rowman & Littlefield, 2019. 107–119.
Tomlin, Liz. *Political Dramaturgies and Theatre Spectatorship: Provocations for Change*. London: Methuen Drama, 2019.
Voigts, Eckart, and Merle Tönnies. "Posthuman Dystopia: Animal Surrealism and Permanent Crisis in Contemporary British Theatre." *Journal of Contemporary Drama in English* 8.2 (2020): 295–312.
Wade, Laura. *Posh: Old Money, New Problems*. London: Oberon Modern Plays, 2010.
Žižek, Slavoj. *Violence: Six Sideways Reflections*. London: Profile Books, 2009.

Matthias Göhrmann
The Spectre of Utopia/Dystopia: The Representation of Anthropogenic Global Climate Change as Culture-War Issue in Richard Bean's *The Heretic* (2011)

Introduction: Global Climate Change and Social Polarisation

Despite the persistency of the COVID-19 pandemic, anthropogenic global climate change is pushing itself back into the spotlights with the brutal urgency of an imminent existential crisis.[1] In October 2021, for instance, just ahead of the UN biodiversity summit in China, business leaders warned of a "dead planet," criticising a lack of the "ambition and specificity required to drive the urgent action needed" in tackling climate change (qtd. in Greenfield).[2] With the language of apocalyptic prophecy, they thus joined progressive governments as well as climate scientists and activists in sounding the shrill alarm over the ecological emergency. In fact, this dramatic language use is predominant when talking about (and thus conceptualising) global climate change. As Rowland Hughes and Pat Wheeler observe, in public debates "climate change is most commonly, and most forcefully, communicated in the language of disaster" (2), heavily relying on negative emotions in order to convey the urgency of the situation and the necessity of comprehensive action. Such *pathos*-driven rhetoric is, however, not unproblematic. In fact, overemphasising, for instance, fear as primary mobilising strategy could prove detrimental to ramping up support for the radical change that is indispensable for averting an incipient climate catastrophe – change that fundamentally depends on challenging deeply held convictions

---

[1] The escalation of the Russo-Ukrainian War in Februrary 2022 has sparked heated debates about Europe's dependency on Russian fossil fuels, underlining the importance of sustainable green energy.
[2] In November 2021, the conclusion of the UN Climate Change Conference in Glasgow, Scotland seemed to reconfirm this assessment, with climate activist Greta Thunberg calling the COP26 "a greenwash campaign, a P.R. campaign" (qtd. in Specia et al.) and Vanessa Nakate remarking that "world leaders have failed to rise to the moment" (qtd. in Harvey).

https://doi.org/10.1515/9783110758252-012

characteristic of the Capitalocene.³ Such rhetoric could backfire and result in disaster fatigue and apathy. It could also fuel a polarising social conflict between those who regard the respective other side of the divide as irreconcilable with their way of seeing themselves and the world, constructing their identities either around the inconvenient truth that is climate change or around some narrative that offers them a less gloomy outlook on the future.⁴ Gregory Garrard seems to share these concerns as he suggests that "political polarisation, not scientific ignorance or deliberate misinformation, is the biggest obstacle to progress on climate change mitigation" (111). And as the divisive debates about Brexit and the COVID-19 pandemic have shown, and British playwright Dennis Kelly astutely observes with regard to dystopian developments within contemporary British culture, the UK is disintegrating "into smaller and smaller groups, focusing on the things that separate us from other people, diving so far into our various identities that we can no longer see each other as humans" (79). It thus appears that in a post-truth world in which scientific consensus is greeted by social division, anthropogenic global climate change has become yet another polarising culture-war issue (Hoffman).

Naturally, contemporary British playwrights have picked up on the predominant atmosphere of eco-anxiety, existential dread and a sense of inescapability

---

3 The term *Capitalocene* refers to capitalism as "world-ecology" (Moore 79) as opposed to world-economy. Jason W. Moore defines it as being based on the Cartesian dualism between body and spirit, that is, nature and culture (83, 87) and maintained through regimes of knowledge and "politics of truth" (86). It is favoured here since this "way of organizing nature" (Moore 6) more clearly puts the emphasis on the epistemological and ontological consequences of such a split as well as the power relations and imbalances that are implicated in the way nature has enlisted for capital. The term *Anthropocene* is comparatively undifferentiated in this respect because it suggests a human universality when it comes to explaining the reasons for global climate change and its effects (Moore 81) and thus uncritically disregards questions of power and the intersectional quality of this crisis. The notion of *Anthropos* upon which it is based, for instance, carries strong classist, racist, Eurocentric and sexist undertones, since it is derived from the notion of the affluent, white, European male. It appears to be necessary to strike a balance between assigning responsibility for the current state of affairs and cooperating to avert the climate change catastrophe. For a detailed discussion of these two terms, see *Anthropocene or Capitalocene?*, edited by Jason W. Moore, and "'Anthropocene, Capitalocene, Chthulucene'" by Anna Grear.

4 As Stanley Cohen's research suggests, denial can be "an unconscious defence mechanism for coping with guilt, anxiety and other disturbing emotions aroused by reality" (5). This soothing of the mind may take the form of an overreliance on technological innovations and their potential to fix climate change, anachronistic behaviour that is based on nostalgia for some romanticised past or climate change denial fuelled by conspiracy theories.

that emanates from the looming prospect of the planet dying.[5] Not only does British climate change theatre thus seamlessly fit in into the larger trend of dramas which deal with (and process) a sense of fear, anxiety and crisis, as has been observed by theatre and literary scholars (Angelaki, Brusberg-Kiermeier et al., Baumbach, Balestrini et al.).[6] Due to their subject matter, eco-plays – like related genres such as climate change fiction (or cli-fi) and eco-poetry – also seem to draw on the "eco-dystopian 'vocabulary'" (Hughes and Wheeler 2) that dominates current debates about anthropogenic global climate change, bringing them close to the genre of dystopia. Indeed, the dystopian turn[7] in British drama has produced eco-plays that employ "the trope of the dystopian near-future" (Reid 72) and/or can be attributed to the subgenre of "absurdist dystopias" (Tönnies 156). There are, however, also more realist and naturalist takes on the psychological and social effects generated by the imminent threat of an ecological catastrophe and the ways in which highly neoliberalised societies make sense of it with more subtle nods to the apparent dystopian qualities of the present. An increasing number of eco-plays, for instance, depict how "individuals [...] must grapple with the public and private dilemmas wrought by climate change" (Johns-Putra 271). In these representatives of British climate change the-

---

[5] After global climate change had been strangely absent from British stages, plays that deal with it in one way or the other have been up-and-coming since 2004: then, Clare Pollard's eco-dystopian *The Weather* featured in the Royal Court's Young Writers Festival, marking the beginning of British playwrights' fascination with the subject matter (Sierz, *Rewriting* 59). Steve Waters' double bill *The Contingency Plan* (Bush Theatre) gave this fascination a well-deserved boost in 2009. And with the multi-authored *Greenland* (National Theatre), Richard Bean's *The Heretic* (Royal Court) and Duncan Macmillan's *Lungs* (Crucible Studio Theatre) the year 2011 alone yielded three British climate change plays, inducing Aleks Sierz to declare that "the issue of climate change now threatens to swamp the programmes of our flagship theatres" ("Heretic"). This trend did continue. Today, climate-change plays belong to a well-established subgenre of British drama, as the following list suggests: Mike Bartlett's *Earthquakes in London* (National Theatre, 2010); Thomas Eccleshare's *Pastoral* (Soho, 2013); Tonya Ronder's *Fuck the Polar Bears* (Bush Theatre, 2015); Emma Adam's *Animals* (Theatre 503, 2015); Stef Smith's *Human Animals* (Royal Court, 2016); Ella Hickson's *Oil* (Almeida Theatre, 2016); Caryl Churchill's *Far Away* and *Escaped Alone* (both Royal Court 2000, 2016); Lucy Kirkwood's *Tinderbox* and *The Children* (Bush Theatre, 2008; Royal Court, 2016) and Clare Duffy's *Arctic Oil* (Traverse Theatre, 2018).
[6] For a concise overview of crisis plays see Dom O'Hanlon's compilation of twenty scenes from contemporary theatre plays which are preceded by short interviews with the respective playwrights.
[7] As Aleks Sierz and Merle Tönnies contend, there are several stages of such a turn: its precursors appeared on British stages in the 1990s. In the 2000s, plays took a turn to absurdist dystopias, before they started mixing realist and absurdist elements in the 2010s (23–25).

atre, it is especially scientists who struggle to relay their findings to family members, friends, politicians and climate activists; that is, laypeople who are characterised by increased scepticism concerning science and/or politics as well as riled by the post-truth rhetoric of disaster and polarisation. In this respect, Steve Waters' *The Contingency Plan* (Bush Theatre, 2009), Mike Bartlett's *Earthquakes in London* (National Theatre, 2010), Richard Bean's *The Heretic* (Royal Court, 2011) and the multi-authored *Greenland* (National Theatre, 2011) are especially noteworthy.[8]

This article focusses on one of these plays, Richard Bean's realist five-act satirical comedy *The Heretic*, and investigates the spectre of utopia/dystopia that haunts it. The analysis is based on the premise that in British society and elsewhere the issue of global climate change is mediated through an increasingly brutalised language. More precisely, this article will examine the play's dramatisation of the culture war, a concept popularised by American sociologist James Davison Hunter. As a collective term, the culture war comprises all the disparate social power struggles over issues of cultural identity which in Britain, too, appear to have become an integral part of political and social life, ultimately amounting to the contestation of the national character and reality itself (Hunter 50, 52). Thus, the culture war constitutes a divisive national discourse which emphasises questions of inclusion and exclusion, sameness and difference, authenticity and spuriousness.[9] It is the culture war's discursive workings, especially its rhetorical qualities, as well as its reliance on neoliberal thinking patterns that contribute to the seemingly unforgiving nature of this social conflict. Generally, the culture-war discourse (mis)represents complex issues as well as multifaceted individuals and groups of individuals as undifferentiated and one-dimensional (Göhrmann 117). This reductive impetus is intensified by the discursive mechanisms of cultural capitalism which, following neoliberal ontology, produces singularised subjects who conceptualise the Self as dependent on the constant competitive performance of uniqueness or a particular aesthetic style so as to feign authenticity (Reckwitz 40). In this way, the culture-war discourse conditions the understanding of the Self and the Other as essentialised subjects and thus shapes the relationship between social actors according to a simplistic friend-or-foe mentality (Göhrmann 122). Eventually, its divisive and reductive "way of knowing" promotes thinking in stereotypes, encouraging an entrenched cultural

---

8 Also consider William Boles and Stephen Bottoms's studies on these plays.
9 For a British Cultural Studies reading of Hunter's concept in the context of UKIP's Brexit campaign, see Göhrmann "Brexit and the Struggle to Define Great Britain." (2021).

warfare between opposed blocs which cannot look past their perceived differences.[10]

The discursive quality of the culture war determines the "battlegrounds" of these social conflicts. Predominantly, hot spots emerge in institutions that produce, regulate and disseminate cultural knowledge into society such as the media, social media networks, schools and universities. There, "skirmishes" ensue between traditionalists and progressives who, rather than constituting monolithic blocs, position themselves at a given moment and with regard to a specific topic by acting upon polarising impulses (Hunter 43–48). In this context, language plays a special role because the culture war devised its own rhetoric of polarisation which is employed by traditionalist and progressive groups alike and is sustained by a decidedly dystopian and utopian imagery which is marked by accusatory name-calling and self-elevating virtue signalling: on the one hand, either side of the debate uses belligerent language as to demonise and to frame the respective Other as an agent of dystopia who, following a rigid orthodoxy, schemes to install a totalitarian system. On the other hand, the Self is considered rather self-righteously and in potentially equally rigid terms as a morally upright deliverer of a pseudo-utopian future, exhibiting the "right" national character (Göhrmann 116).

This article therefore focuses on the representation of the setting as well as the characters, that is, their behaviour and language use. With the definition of the culture war in mind, reading the ways in which *The Heretic* dramatises a British society that has become increasingly neoliberalised, fragmented and self-involved reveals the restricting and damaging nature of the black-and-white thinking that implicates the perception of issues, and in turn individuals, that in fact do have a lot of grey areas. More concretely, this article contends that the dystopian as well as utopian qualities of the play, subtle as they might be, are reflected in the way Bean depicts the treatment of *The Heretic*'s main character, Diane, and how she defies a system represented as increasingly suppressing deviant opinions and even facts. In the face of this orthodoxy and the divisive rhetoric dominating the climate change debate, the ending of the play is read as a satir-

---

10 The emergence of the culture-war discourse and its rhetorical strategies rests upon developments which have taken place since the 1960s and have become more prevalent ever since, two of which are of particular importance here: first, the replacement of "class" by social identity as the primary means of identification. Second, the growing awareness in post-industrial societies that reality is linguistically and discursively mediated. These processes have yielded "identity politics" and the "culturing of politics" respectively (Hall, "Pathways" 167), which may facilitate attempts at stereotyping; that is, fixing and naturalising difference (Hall, "Spectacle" 247). Both kinds of politics have merged with the competitive rationale of neoliberalism.

ical take on the utopian yearning for a romanticised "normalcy" in which the polarisation over questions of identity is superficially dismissed but eventually remains unresolved.

"Believers" vs. "Deniers": The Post-Truth Politics of a Green Orthodoxy?

The *Oxford English Dictionary* defines a heretic as someone "who maintains theological or religious opinions at variance with the 'catholic' or orthodox doctrine of the Christian Church," also providing the reader with compounds such as "heretic-hunting" or "heretic-burning" which give some insight into what the consequences of being such a non-conformist might have been (*OED* s.v. "Heretic"). The title of Bean's play therefore perfectly encapsulates its argumentative claim, namely, that the media, politicians as well as climate activists and scientists engage in the topic of anthropogenic climate change with such religious zeal and rigid morality that a slight deviation from or criticism of the authoritative narrative of climate change equals a (social) death sentence.[11] The play's conflicts therefore revolve around the eponymous heretic, Dr Diane Cassell: rigorously following her empirical research, the sea-level expert at the University of York finds that, despite all predictions and apocalyptic prophecies, sea levels in the Maldives have not been rising for the past 16 years (22).[12] Her findings therefore contradict the majority of research on rising sea levels, which is also presented as the overwhelming scientific consensus in the play, and if they do not fully deny the probability, then they at least weaken the prospect of global ecological

---

[11] It is interesting that with his focus on climate change Bean chooses to address a culture-war issue which is arguably not as much of a hot-button topic as discrimination based on "race," gender and sexual orientation. In this context, his criticism of religious orthodoxy is also somewhat defused in its explosiveness by comparing it to climate change activism – and this in a satirical-comedic fashion. In this way, Bean perhaps aims at avoiding a backlash about some of the play's controversial presumptions.

[12] This is, in fact, pure fiction: as the Maldivian Minster of Environment, Climate Change and Technology Aminath Shauna states, 90 per cent of the country's islands have already experienced flooding and 97 per cent of them have reported shoreline erosion (Gilchrist). The situation is especially precarious for the South Asian archipelago, since 80 per cent of the country's islands are only one metre above sea level and scientists predict a sea level rise of up to 1.1 metres by 2100 (Gilchrist). The Maldives therefore dramatically symbolise the very urgency for action against anthropogenic global climate change, which might have induced Bean to have Diane conduct her research in that country. In *The Heretic*, these facts are disputed by her in an interview with Jeremy Paxman and the Maldives High Commissioner in London (43–45).

catastrophe. Following the principles of debate-based drama, the play is set in a realistic present-day Britain and depicts characters who, in the ensuing debate, occupy "dialectically opposed positions" (Bottoms 340).[13] More precisely, there are two central conflicts which arise from Diane's research and set her against two distinct "warring parties": the first is an argument of and about economic interests and takes place between Diane and her ex-lover and superior Professor Kevin Maloney, who is the Head of Faculty Earth Sciences at the University of York. The second is a clash of and between mentalities which expresses itself in the science-religion binary. Here, Diane is at odds with her 21-year-old anorexic Greenpeace-member[14] daughter Phoebe, her 19-year-old student and climate activist Ben Shotter, and the security officer Geoff, who – as it turns out – is the only member of a radical climate activism "group" called the Sacred Earth Militia and threatens to murder Diane for her heresy. Both conflicts are framed by a decidedly post-factual politics of truth that is emblematic of the culture war, with Diane's antagonists seeking to exploit scientific research for either neoliberal profit maximisation or an oppressive green orthodoxy.

The play's assessment of the culture war's neoliberal qualities is particularly discernible in the first half of the play (acts one to three), more precisely in its

---

13 As Stephen Bottoms notes, "[b]inarized dramatic structures of this sort have traditionally been used to explore the political spaces between left and right, or between moderate and radical" (340). Due to the culture war, this divide is now "culturalised." *The Heretic* follows this binary setup and depicts a social conflict between characters who appear to personify the opposing sides in an almost stereotypical fashion: there is, for instance, Phoebe, the "anorexic Greenpeace daughter" or Kevin, the "tweedily corrupt professor" (Pearce). In general, there appears to be a return to "stock characters" across genres, be it in theatre plays, novels or political rhetoric. In this way, cultural products of our time process the social polarisation and the pervasive cultural divisions that dominate the public as well as private sphere, as has become obvious in the debates surrounding Brexit and COVID-19. These tendencies are read here as results of the reductive force of the culture-war discourse.

14 In the play, Greenpeace, the WWF or the Intergovernmental Panel on Climate Change (IPCC) are considered political pressure groups that do politics and not science. Politics is seen here as an extremely corrupting element that in its pursuit of power and influence bends science to its will. This depiction is due to, amongst other things, *The Heretic* being the dramatic response to two connected scandals that have undermined the public's trust in climate science. As Stephen Bottoms points out, the political disenchantment which is discernible in the play is based on the "notorious factual inaccuracy in the IPCC's Fourth Assessment Report" in 2007 (343). Like Kevin in scenes not discussed in this paper, its author, Dr Murai Lal, admitted later that he had deliberately exaggerated the prospect of glaciers melting to provoke a political response (Bottoms 342–343). Lal only confessed, however, after investigative journalists had scanned through a host of email correspondence between researchers at the University of East Anglia which had been hacked by climate change sceptics in 2009 (343). In an attempt to frame anthropogenic climate change as a fabricated hoax, deniers soon dubbed this scandal "Climategate."

setting and the character of Professor Kevin Maloney. Here, the action is set in a "modern university head of department office" (9), emphasising the central role of universities in the culture war. Bean notes that the furniture's "dull functional designs" should make it clear to the audience that "this is not Oxbridge" (9). The modern appearance of the setting is read here as echoing Blair's rhetoric about his project of modernisation in the 1990s which aimed at preparing "New" Labour and Britain for meeting the "new challenges" of the "new millennium" (Fairclough 18–19).[15] In fact, Blair's tendency "to reduce national identity to a corporate brand" (Parekh 261) perfectly catches the problematic state of contemporary Britain which informs *The Heretic* and is arguably foreshadowed by the description of the setting. Sticking to Blair's project for a moment, the word "modern" has two connotations: on the one hand, it describes a more inclusive and tolerant agenda that defined "New" Labour as subscribing more comfortably to the ideas of diversity, multiculturalism and social justice (Parekh 261). In this context, not setting the action at Oxbridge more clearly emphasises the progressive character of this public university and the values it represents. On the other hand, however, Blair propagated the idea that Britain would only find its way into the future if it took the so-called "Third Way." The second connotation of the word "modern" therefore references the continued neoliberalisation of British society as well as its state and institutions. In these two senses, then, the modern setting is considered a visual cue that alludes to the background against which *The Heretic* is to unfold: a public university turned knowledge production factory which has succumbed to the principles of economic competition and profit maximisation, constituting the central "battleground" for the discursive contestation of "truth" in the larger context of a post-industrial society that is in the conscious process of renegotiating its cultural values.

That neoliberal ideology is indeed deeply ingrained in this institution of higher education – that is, in the thinking and practices of most of its teaching and administrative staff – becomes apparent throughout the play in the way some of the characters behave and act. Whereas in this respect Diane again does not conform to hegemonic thinking patterns, her direct superior is all for the neoliberal approach to education and research: Kevin repeatedly barges into Diane's office, confronting her, for instance, over still giving one-on-one tutorials, since that is not a cost-efficient way of teaching (52). His attempts at micro-managing his sea level expert in a rather patronising way clearly follow

---

15 This interpretation is, of course, much easier to follow when reading the dramatic text rather than watching the play's theatrical performance, since the secondary text itself explicitly mentions "modern" and the "dull functional designs" of the furniture (9). This may look different in the actual production of the play.

strategic considerations that originate from the monetary incentives of capitalism and its principle of profit maximisation. This becomes especially apparent in Kevin's language use: he is certain, for instance, that his department is in need of "a new 'business model'" (22), that a prospective "'Climate Change Research Unit' would service clients" (23) and that he and Diane "are the Earth Sciences Faculty of YUIST for teaching, but *virtually* [they] are a separate budget centre, providing tools to the market" (23). This neoliberal lingo thus represents the pervasive force with which universities have been corrupted by the system of equivalency that seeks to assign a market value to research – a development that has increased competition between disciplines, departments and universities, as Kevin points out at the end of act one, scene two (24–25). More precisely, it is the race between the University of York and the University of Hampshire for investments from one of Europe's biggest insurance and underwriting firm, Catalan International Securities, which motivates Kevin's actions (22). Both universities are locked in a struggle for financial means which corrupts the integrity and undermines the independence of scientific research. Since Kevin's department has to compete with departments from other universities whose publications are all in line with the dominant narrative of an imminent ecological catastrophe, Diane's findings have the potential of jeopardising his financial aspirations which depend on the conformist image of his department. This is why he asks her to delay the publication of her research until after the Catalan decision has been made (23–24).

Against the backdrop of this neoliberal setting, the conflict between empirical evidence and what belongs to the realm of make-believe, wishful thinking or deliberate untruths makes climate change and the reality it creates an increasingly slippery concept. Whilst the difficulty of representing climate change (its significance, effects and meaning) on stage is one of climate change theatre's main challenges, the post-factual breakdown of meaning and the contestation of established "truths" in *The Heretic* is read here as an allusion to the culture-war discourse and its tendency to depict issues, people and events in a simplistic either-or fashion. This is particularly noticeable on the language level where the inclination to employ hyperbolic as well as denunciative rhetoric to polarising effect is especially prominent in the younger generation, as represented by Diane's student Ben Shotter and Diane's daughter Phoebe.

> BEN: You're a denier. Right?
> DIANE: Holocaust or –
> BEN: (*Sighing.*) – Tut. *Anthropogenic global warming.*
> DIANE: I'm agnostic on *AGW*, but if you can prove to me
>   there's a God I'll become a nun quicker than you can say
>   'lesbian convent orgy.'

BEN: That's sexist.
DIANE: No.
BEN: Homophobic.
DIANE: Yes. Semantic specificity is the E chord of science.
BEN: Are you allowed to be homophobic in a government funded educational establishment, in the, like, you know, public sector?
DIANE: I doubt it. Why did you choose Earth Sciences? (26)

Without a doubt, Diane's sarcastic retorts produce much laughter and contribute to the satiric quality of the play. Although she employs sarcasm to teach her student semantic specificity, she also uses it as strategy to maintain her position as a researcher: labels such as climate "change denier," "Holocaust denier," "sexist" and "homophobe" mark the socially unacceptable and tabooed and could be harmful to Diane's social and professional standing. These attempts at discrediting Diane by assigning unfavourable labels to her are typical of the *ad hominem* reasoning facilitated by the culture-war discourse's reductive impetus. They also show the polarising and rather undifferentiated ways in which anthropogenic climate change is understood: Diane does not have to be a fully-fledged denier, that is, someone who claims global warming to be a hoax and categorically rejects all research and every scientific principle on the subject matter.[16] For being stereotyped as "denier" it suffices for her to disagree with very few aspects of the predominant narrative about climate change.[17] Clearly, Bean bases this aspect of his play on the self-victimisation employed by individuals who vocally and

---

[16] Although Diane is sceptical of computer models that predict an apocalyptic future, she reveals her allegiance to most of the science behind global climate change when she is put through her paces by Kevin (35–36).

[17] This obviously takes away from the complexity of the issue. Probably, there could be climate change and at the same time still no rising sea levels in the Maldives; there can certainly be global warming and snow in winter. As Kevin himself so fittingly puts it in his retelling of a row with Gordon Brown: "I said 'Ah! Not my fault, Prime Minister, I'm your *climate* expert, that *snow* is weather.'" (21) And also on the level of visual cues, Bean seems to play with the confusion between climate and weather which is so often present in arguments of "deniers" by suggesting an ordinary passing of seasons in the secondary text: Act one starts in September. In act two "[i]t's colder now, and GEOFF is wearing a high vis coat. KEVIN is also wearing a winter coat." (31). In act three "[h]eavy snow is falling" (46). Act four starts on Boxing Day and the audience is to be made aware of the fact that "although it is not snowing now, snow has fallen and drifted" (57). The play ends in August with "[s]unshine, birdsong" and "[t]he door open, with summer sun streaming on the kitchen floor" (92). However, this could also be read as nature symbolising the inner state of and perhaps even the relationships between the characters – an anthropocentric appropriation of "nature" that more progressive (or unconventional) eco-plays intend to question.

media-effectively criticise the prospect of being "cancelled" due to their violation of some perceived Orwellian speech and/or thought protocol. But in a more fundamental way, he alludes to the erosion of the middle ground on which debates can take place: Diane's reasoned criticism of one aspect of a hegemonic narrative is silenced, not by engaging with her research through constructive debate, but by shunning her and attacking her personally, leaving her with no other way to respond than brushing it off with biting sarcasm. Efforts at policing Diane's behaviour by way of accusation and contrastive virtue signalling occur several times in the play. This happens, for instance, when Phoebe disagrees with her mother and calls her a fascist (11, 84), or "a gas guzzling planet rapist," whilst emphasising that she herself is a "member of Greenpeace" (59), that is, a good person. Ben's attempts at denouncing Diane due to her disregard of politically correct language persist, although in a rather playful way (46).

It is curious that even though Ben keeps admonishing Diane, he continues to attend her tutorials. Besides learning about statistics and how politics, activism and science do not go well together, he works with Diane to uncover a conspiracy involving the University of Hampshire's leading climate change expert, whose figures simply do not add up.[18] In these meetings Ben showcases his personal moral standards, his quick-to-judge attitude and a general moral rigidity.[19] These are illustrated, for instance, by his dismissal of Einstein's work on grounds of the theoretical physicist having cheated on his wife, which prompts Ben to call Einstein a "wanker" (46), and allusions to Ben's relationship with his father. As it turns out, he despises his father simply because he drives a Volvo (28) and because a year ago, when Ben was eighteen, he caught his father watching porn, inciting him to hit his father with a casserole dish (49). Confronted with the news that his father has actually changed his ways Ben experiences cognitive dissonance, confessing to Diane that "[s]omething terrible has happened. [...] My dad has gone green. Like bare mad green. He's scrapped the Volvo, bought solar panels, started cycling. And my brain is like totally fing!" (49).[20] Naturally,

---

18 This project makes Diane and Kevin allies. Even though they have different motivations – Diane seeks to disprove faulty scientific assessments, Kevin to obtain funding – they effectively attempt to undermine the standing of the University of Hampshire. Arguably, it is this alliance which contributes to the play's exaggerated harmonious ending.
19 It should be noted that in the overall development of the play's action, Diane interviewing Ben about his aggressive eco-morality clearly serves to keep up suspense: at this point the audience do not know yet who the Sacred Earth Militia are and who wrote the death threats against Diane and are tempted to think it was Ben.
20 Ben's allegedly uncompromising morals also express themselves in his critique of extramarital sex and masturbation as well as his abstinence which is reminiscent of the tendency of re-

this "loveably gormless Ben" (Sierz, "Heretic") evokes many laughs and contributes to the comedic quality of the play. However, inherent in this character is Bean's satiric criticism of the ways in which a young generation constructs their identity around dualist thinking patterns and how such thinking – based on the essentialised difference between the Self and the Other – has resulted in the unforgiving unwillingness (or perhaps inability) to look past perceived moral transgressions or even entertain the thought that the Other is a multifaceted fellow human being. This categorical dismissal of the Other is an important aspect of the polarising force of the culture-war discourse, glossing over nuances and undermining any compromise: instead of seeing his father going "bare mad green" (49) as an opportunity to make up with him, Ben looks for ways that allow him to continue hating him out of principle (49–50). Bean's persiflage of adolescents who consider and advertise themselves as progressive and tolerant but turn out to be orthodox and intolerant is also visible in the way he escalates Ben's climate activism, alluding to the ways in which a younger generation takes as well as demands radical individual responsibility for averting a global climate catastrophe. The play starts off with Ben informing Diane about missing class because he rejects going on a school trip by minibus – he prefers turning it into an 80-mile bike ride instead (12–13). In another meeting with Diane, he confesses to having moved off campus onto a barge because over the colder winter months the student residences "turned the heating on" (43). Ben's activism reaches an absurd climax when he takes a rather misanthropic stance and advertises the Voluntary Human Extinction Movement which works "towards removing human life from earth by non-reproduction" (48).[21] Bean then has Ben profess to Diane that he only eats locally grown vegetables and aims to reduce his "emissions" of methane and CO_2 that are caused by flatulence and breathing by eating garlic in large quantities (48). Besides poking fun at the stereotypical "gassy, smelly vegetarian," Bean's depiction of the self-sacrificial fashion in which Ben takes it upon himself to improve the planet also satirises the ways in which capitalist realism makes individualised responsibility for saving the planet appear commonsensical. Thus, it is – to borrow the title of Joss Sheldon's dystopian novel – the narrative of a neoliberal *Individutopia* that dis-

---

ligious groups to regulate sex and oppose sexual liberation movements. Bean's representation of climate activists therefore emphasises (or even ridicules) their highly moralist stance.

21 Note that Bean reveals Ben's hypocrisy by making him betray his principles. Ben could, for instance, chime in and cease all contact with Diane since she is branded a denier, but he does not. And in the last act, too, Ben has partly changed his ways: he still prefers riding a bike over taking a ride in a car. But he was apparently less strict with his non-reproduction policy, since his wife-to-be, Phoebe, is pregnant.

perses all efforts for a radical social restructuring of society and economic redistribution of wealth – steps which are most likely necessary if anthropogenic global climate change is to be tackled successfully.[22]

To be sure, Diane also does her fair share of stereotyping: she assumes that Ben is unfit for becoming a scientist due to his unconventional clothing style, psychological issues and his climate change activism, concluding that he is best suited for art college (28). And towards the end of the play, she insists in the language of dramatic hyperbole that "[c]ars are liberating, democratic, and feminist. And the day when Greenpeace has succeeded in pricing the poor off the roads will not be a good day for the planet, it will be a good day for totalitarianism" (80). The Other, Greenpeace, is stereotyped here as agent of dystopia which schemes to install an ecologically friendly, albeit totalitarian system. Here, Bean alludes to the class bias in discussions about how to tackle climate change, with ostensibly affluent members of society who can afford to indulge in a green lifestyle, supporting "some expensive politically correct green act of 'bunny hugging'" (Boris Johnson qtd. by Guardian Staff) whilst turning a blind eye to the restrictions of financially disadvantaged social groups. In this instance, he also depicts the impulsive language of accusation so prominent in present-day political discourse and the rhetoric of polarisation of the culture war. In *The Heretic*, too, this clearly obstructs any rational debate on the matter at hand, making Diane's pride in the scientific method and her behaviour, which is at times self-righteous, stubborn and condescending, contributing factors to the conflicts between her and on the one hand, Kevin and on the other, Phoebe and Ben. Whereas Ben's attempts at "branding" Diane a non-conformist on grounds of her insensitive language use are futile, her stigmatisation as a climate change denier does not remain without consequences.

Because of her insistence on her professional convictions as well as on publishing her research findings, Diane's career is in jeopardy and her life in danger. She receives three death threats from Geoff's Sacred Earth Militia (16, 31, 79) who considers Diane a "heretic" (17) – threats which are downplayed by Kevin, who prefers to focus on business plans, ways of maximising profit and getting to have sex with Diane. Following the traditional Aristotelian structure of five-act drama,

---

[22] As Mark Fisher points out, the radical change that is needed to mitigate anthropogenic climate change is eventually obstructed by capitalism's very own nature and the way it has been naturalised as a way of living, being and knowing without alternatives (4). The consequences of capitalist realism appear obvious: solutions to the problem at hand are framed by a decidedly neoliberal outlook that propagates – besides the individualisation of responsibility – technological progress, monetary incentives for going green or some economic "green growth" as preferred answers to ecological collapse.

the conflict between Diane and Kevin climaxes in act three and results in her suspension, making it ever more probable that the end of her career is imminent. It is through the prospective tragic fate of Bean's proud scientist that *The Heretic* comments on the perceived fanatical zeal with which some activists demand the radical reorganisation of social, political and economic structures in Britain and the West by allegedly curtailing the freedom of expression.[23] In Bean's Britain, research findings and empirical data that contradict the scientific consensus on global warming are not discussed, but seen as treacherous heresy, whilst unwavering loyalty to the hegemonic narrative about climate change is demanded, perverting science into an oppressive quasi-religion – the dystopian trope of science exploited by a totalitarian regime becomes obvious here. "Religious" belief and scientific fact are equated with each other, with both sides of the divide thinking in a similarly dismissive fashion about the respective opposition. For Diane, climate-change activism is "the perfect religion for the narcissistic age" which is – like any other religion – "shot through with inconsistences" (80).[24] And for Phoebe "[e]mpiricism is a fucking ism like any other fucking ism!;" she admonishes her mother to "[s]top evangelising [her] lousy religion!" (87).

Conclusion and Outlook: Fighting about Nature, Fighting about Culture

In the end, however, all's well that ends well: Diane's research is more or less trusted, and even though Phoebe gets worked up about her mother's convictions – and in a retarding moment of the play suffers a major heart attack in the process (87–91) –, *The Heretic* closes on a reconciliatory note that could be described as a truce between the "warring parties." Kevin offers to re-employ Diane (83), Phoebe is rescued by the Sacred Earth Militia / Geoff (90–91) and Ben and Phoebe, who have fallen in love, are expecting a child and are about

---

[23] In the play, this is emphasised by the reference to another professor who received death threats from Islamic fundamentalists due to pointing out grammatical errors in the Quran (18) – yet again, a mix of banality and a disproportionately aggressive reaction.

[24] This equation of climate activism and religion is foreshadowed in the first scene of the first act when Phoebe compares Ben's stance on riding on a minibus to religious convictions: "Muslims and Jews can't eat pork. He can't go on a fossil fuels minibus" (12). When Diane shows herself unimpressed by this reasoning, her daughter calls her out on her alleged "religious discrimination" (13).

to get married (92–95).²⁵ With the shift of focus away from the prospect of a global climate catastrophe, Phoebe surviving and Diane's rehearsal of her wedding-speech at the end of the play, *The Heretic* suggests a return to a romanticised normalcy that, compared with the apocalyptic language used to impart the very realistic prospect of ecological collapse in the real world, could appear to some audience members as a pseudo-utopian present where "there is really nothing to worry about" (Bottoms 343). Although – or perhaps because – this satirical comedy certainly has produced much laughter (Sierz, "Heretic") and "ends in a warm glow of humanist sentiment" (Billington), *The Heretic* does not resolve any of its central conflicts. With their competitor's fraudulence exposed, the question about Diane's now less sensational "heresy" fades into the background, Kevin's aspirations for a new research unit seem feasible and departmental domination within the University of York consolidated.²⁶ Or to phrase it more concisely: neoliberalism's grasp on research for the sake of profit maximisation has not weakened and is still presented as being without alternative. And the second conflict, too, is far from settled: the way in which "knowledge" and "fact" on the one hand, and "belief" and "opinion" on the other, continue to be understood as having the same argumentative value appears as an unresolvable dilemma which a young, angry and yet idealistic generation – one which has internalised the urge to polarise and singularise – seems ill-equipped to solve alone.

Bean's flirtations with climate-change scepticism are provocative but prove stimulating nonetheless. The criticism of the culture war and its polarising discursive workings is inherent in *The Heretic*, as well as the frustrated hope of listening to the opposition and not dismissing it right away so as to make debates about controversial issues possible at all. In the play, this is mostly achieved through Bean's treatment of the "heretical" yet proud scientist Diane. Speaking about Bean's veneration of the play's main character, Michael Billington aptly asks, "[w]ould he extend the same charity […] to a flat-earth advocate?" And one must indeed wonder whether there are topics which are off-limits, and if so, how one is to reach a consensus on what those are – and if not, how one

---

25 There is probably no better way of symbolising the mutual respect for differing opinions than Kevin planning the trip to the wedding venue at the end of the play: "Right! It's me and the bride in the Jag, with the top down. You're driving the Prius, and Ben is on his bike" (93).
26 Bean's problematic reinterpretation of the "Climategate" scandal at the University of East Anglia in 2009 provides the *deus ex machina* solution to Kevin and Diane's dispute. A friend of Ben's hacks into the University of Hampshire's email servers, confirming what Diane had suspected all along: Hampshire's leading climate change researcher manipulated his findings to fit in with the global climate change hysteria. With their competitor discredited, the University of York will most likely convince Catalan to invest in their Earth Science Department.

could discuss them in more composed ways. Real world examples of this predicament abound. The central culture-war conflict in *The Heretic* – that is, the way the truth about global climate change is discursively constructed and contested in a polarising manner –, for instance, is eerily mirrored by what has been happening at the University of Sussex. In January 2021, philosophy Professor Kathleen Stock was criticised by "professional academic philosophers committed to the inclusion and acceptance of trans and gender non-conforming people" in an open letter for her "trans-exclusionary public and academic discourse on sex and gender" ("Open Letter"). The dispute gained political explosiveness when the Conservative government awarded Stock an OBE, thereby demonstrating the kind of political instrumentalisation of academia in the overall struggle about the interpretative hegemony over "reality" which is repeatedly criticised by Diane in the play (Bean 26–28). Back then, Stock called "for UK universities to end their association with Stonewall, the prominent LGBTQ+ rights charity, describing its trans-inclusive stance as a threat to free speech" ("Open Letter"). In early October 2021, then, an anonymous group of students called for Stock's termination due to her views on gender identity (Badshah). Three weeks after protests on campus and on social media networks had started, Stock announced her resignation from the University of Sussex (Adams).

There are several hard questions with which *The Heretic* leaves the audience, the most acute of which might be how the fragmentation of (British) society can be counteracted and a respectful democratic debate established – one which integrates disparate perspectives and a multiplicity of identities without succumbing to ideological appeals to a "common humanity" which gloss over the relations of power that in the context of global climate change, have brought Earth's ecosystem to the brink of collapse in the first place. In the end, one must wonder if such a bold democratic endeavour is at all possible. With the culture war drawing its polarising power largely from the hegemony of neoliberalism, cultural capitalism continues to appear to many as a system without any alternative.

Works Cited

Adam, Emma. *Animals*. London: Oberon, 2015.
Adams, Richard. "Sussex Professor Resigns after Transgender Rights Row." *The Guardian*, 28 Oct. 2021. Web. 20 Dec. 2021. <https://www.theguardian.com/world/2021/oct/28/sussex-professor-kathleen-stock-resigns-after-transgender-rights-row>.
Angelaki, Vicky. *Social and Political Theatre in 21st-Century Britain: Staging Crisis*. London: Bloomsbury, 2017.

Badshah, Nadeem. "University Defends 'Academic Freedoms' After Calls to Sack Professor." *The Guardian*, 7 Oct. 2021. Web. 8 Oct. 2021. <https://www.theguardian.com/education/2021/oct/07/university-defends-academic-freedoms-after-calls-to-sack-professor>.

Balestrini, Nassim et al. "Theater of Crisis: Contemporary Aesthetic Responses to a Cross-Sectional Condition – An Introduction." *JCDE* 8.1 (2020): 2–15.

Bartlett, Mike. *Earthquakes in London*. London: Methuen Drama, 2010.

Baumbach, Sibylle. "The Fascination with Crisis and the Crisis of Perception in Contemporary British Drama." *JCDE* 8.1 (2020): 47–64.

Bean, Richard. *The Heretic*. London: Oberon, 2015.

Billington, Michael. "The Heretic – Review." *The Guardian*, 11 Feb. 2011. Web. 7 Nov. 2021. <https://www.theguardian.com/stage/2011/feb/11/the-heretic-review>.

Boles, William C. "The Science and Politics of Climate Change in Steve Water's *The Contingency Plan*." *JCDE* 7.1 (2019): 107–122.

Bottoms, Stephen. "Climate Change 'Science' on the London Stage." *WIREs Climate Change* 3 (2012): 339–348.

Brusberg-Kiermeier, Stefani et al. "Fear and Anxiety in Contemporary Drama and Performance: An Introduction." *JCDE* 7.1 (2019): 1–11.

Buffini, Moira, et al. *Greenland*. London: Faber and Faber, 2011.

Churchill, Caryl. *Far Away*. New York: TCG, 2011.

Churchill, Caryl. *Escaped Alone*. London: Nick Hern Books, 2017.

Cohen, Stanley. *States of Denial: Knowing About Atrocities and Suffering*. Cambridge: Polity Press. 2010.

Duffy, Clare. *Artic Oil*. London: Oberon Modern Plays, 2018.

Eccleshare, Thomas. *Pastoral*. London: Oberon Modern Plays, 2013.

Fairclough, Norman. *New Labour, New Language?* London: Routledge, 2010.

Fisher, Mark. *Capitalist Realism: Is There no Alternative?* Winchester: Zero Books, 2009.

Garrard, Gregory. "Brexit Ecocriticism." *Green Letters* 24.1 (2020): 110–124.

Gilchrist, Karen. "'There's No Higher Ground For Us': Maldives' Environment Minister Says Country Risks Disappearing." *CNBC*, 18 May 2021. Web. 7 Nov. 2021. <https://www.cnbc.com/2021/05/19/maldives-calls-for-urgent-action-to-end-climate-change-sea-level-rise.html>.

Göhrmann, Matthias. "Brexit and the Struggle to Define Great Britain: The Culture-War Discourse and UKIP's Weaponisation of Racialised Identities." *Mentalities and Materialities*. Ed. Philip Jacobi and Anette Pankratz. Würzburg: Königshausen & Neumann, 2021. 111–127.

Grear, Anna. "'Anthropocene, Capitolacene, Chthulucene': Re-encountering Environmental Law and Its 'Subject' with Haraway and New Materialism." *Environmental Law and Governance for the Anthropocene*. Ed. Louis J. Kotzé. Oxford: Hart Publishing, 2017.

Greenfield, Patrick. "Halt Destruction of Nature or Risk 'Dead Planet,' Leading Businesses Warn." *The Guardian*, 11 Oct. 2021. Web. 12 Nov. 2021. <https://www.theguardian.com/environment/2021/oct/11/halt-destruction-of-nature-or-risk-dead-planet-leading-businesses-warn-aoe>.

Guardian Staff. "Greta Thunberg Dubs Herself a 'Bunny-Hugger' after Boris Johnson's Climate Remarks." *The Guardian*, 23 Apr. 2021. Web. 12 Nov. 2021. <https://www.theguardian.com/environment/2021/apr/22/greta-thunberg-bunny-hugger-boris-johnson-twitter>.

Hall, Stuart, ed. "The Spectacle of the Other." *Representation*. Nixon. London: Sage, 2013.

Hall, Stuart. "Some Politically Incorrect Pathways Through PC." *The War of the Words*. London: Virago Press, 1994. 164–183.
Harvey, Fiona. "'The Pressure for Change is Building': Reactions to the Glasgow Climate Pact." *The Guardian*, 14 Nov. 2021. Web. 20 Dec. 2021. <https://www.theguardian.com/environment/2021/nov/14/the-pressure-for-change-is-building-reactions-to-the-glasgow-climate-pact>.
Hickson, Ella. *Oil*. London: Nick Hern Books, 2016.
Hoffman, A. J. "Climate Science as Culture War." *Stanford Social Innovation Review* 10.4 (2012): 30–37. Web. 6 Sept. 2021. <https://doi.org/10.48558/YY4T-Y622>.
Hughes, Rowland and Pat Wheeler. "Introduction Eco-Dystopias: Nature and the Dystopian Imagination." *Critical Survey* 25.2 (2013): 1–6.
Hunter, James Davison. *Culture Wars: The Struggle to Define America*. New York: Basic Books, 1991.
Johns-Putra, Adeline. "Climate Change in Literature and Literary Studies: From Cli-Fi, Climate Change Theater and Ecopoetry to Ecocriticism and Climate Change Criticism." *WIREs Climate Change* 7.2 (2016): 266–282.
Kelly, Dennis. "Dystopia." *JCDE* 9.1 (2021): 77–80.
Kirkwood, Lucy. *The Children*. London: Nick Hern Books, 2016.
Kirkwood, Lucy. *Tinderbox*. London: Nick Hern Books, 2018.
Macmillan, Duncan. *Lungs*. London: Oberon Modern Plays, 2020.
Moore, Jason W. "Introduction: Anthropocene or Capitalocene? Nature, History, and the Crisis of Capitalism." *Anthropocene or Capitalocene? Nature, History, and the Crisis of Capitalism*. Ed. Jason W. Moore. Oakland: PM Press, 2016. 1–12.
Moore, Jason W. "The Rise of Cheap Nature." *Anthropocene or Capitalocene? Nature, History, and the Crisis of Capitalism*. Ed. Jason W. Moore. Oakland: PM Press, 2016. 78–115.
Moore, Jason W. Ed. Anthropocene or Capitalocene? Nature, History, Oakland: PM Press, 2016.
O'Hanlon, Dom. *Theatre in Times of Crisis: Twenty Scenes for the Stage in Troubled Times*. London: Methuen Drama, 2020.
"Open Letter Concerning Transphobia in Philosophy." Jan. 2021. *Philosophy Transphobia Letter* 8. Web. 15 Nov. 2021. <https://sites.google.com/view/trans-phil-letter>.
Oxford English Dictionary. "Heretic." *OED*. Web. 15 Oct. 2021. <https://www.oed.com/view/Entry/86197?redirectedFrom=heretic#eid>.
Parekh, Bhikhu. "Defining British National Identity." *The Political Quarterly* 80 (2019): 251–262.
Pearce, Fred. "Take Climate Scientists to Task, but Avoid Formulaic Boffin-Bashing." *The Guardian*, 11 Feb. 2011. Web. 7 Nov. 2021. <https://www.theguardian.com/environment/2011/feb/11/the-heretic-climate-change-review>.
Pollard, Clare. *The Weather*. London: Faber and Faber, 2004.
Reckwitz, Andreas. *The Society of Singularities*. Cambridge: Polity Press, 2020.
Reid, Trish. "The Dystopian Near-Future in Contemporary British Drama." *JCDE* 7.1 (2020): 72–88.
Ronder, Tanya. *Fuck the Polar Bears*. London: Nick Hern Books, 2015.
Sheldon, Joss. *Individutopia. There Is No Such Thing as Society*. London: Rebel Books, 2018.

Sierz, Aleks and Merle Tönnies. "'Who's Going to Mobilise Darkness and Silence?' The Construction of Dystopian Spaces in Contemporary British Drama." *JCDE* 9.1 (2021): 20–42.

Sierz, Aleks. "The Heretic, Royal Court." *Aleks Sierz: New Writing for the British Stage*, 10 Feb. 2011. Web. Nov. 2021. <https://www.sierz.co.uk/reviews/heretic-royal-court/>.

Sierz, Aleks. *Rewriting the Nation. British Theatre Today*. London: Methuen Drama, 2011.

Specia, Megan, et al. "Greta Thunberg Assails World Leaders for 'Profiting From This Destructive System.'" *The New York Times*, 5 Nov. 2021. Web. 10 Dec. 2021. <https://www.nytimes.com/2021/11/05/world/europe/cop26-greta-thunberg.html>.

Smith, Stef. *Human Animals*. London: Nick Hern Books, 2016.

Tönnies, Merle. "The Immobility of Power in British Political Theatre after 2000: Absurdist Dystopias." *JCDE* 5.1 (2017): 156–172.

Waters, Steve. *The Contingency Plan: On the Beach and Resilience*. London: Nick Hern Books, 2010.

Leila Michelle Vaziri
"I Am the Abyss into Which People Dread to Fall": Encountering Anxiety in Dystopian Drama

In the recent past, there has been no lack in potential sources of anxiety: global heating, Brexit, the war in Ukraine with its renewal of the threat of nuclear conflict, the COVID-19 pandemic and the insurrection at the Capitol have all shaken up certainties that most in the global north had taken for granted. Public reaction to these developments, in street protests and in the media, betrays a sense of anxiety that suffuses social and political discourse. This anxiety is also increasingly manifesting itself in contemporary drama. Indeed, it seems that the last years have seen the rise of what might be called a "theatre of anxiety," which uses an overwhelming sense of anxiety as the backdrop against which topics of social, political, technological and ecological importance are staged and combines them with philosophical and aesthetic implications of anxiety.

Given this central role of anxiety in the current academic and political debate, my aim is to show how this sensation influences and is represented in dystopian drama on both a thematic and aesthetic level. As the following interpretation will show, a particular focus in analysing theatre and anxiety must fall on the staging of time and pain. Both time and pain are transgressive phenomena. The construction of pain as crossing various borders and time as connecting past, present and future can frequently be traced in dramatic representations of anxiety. I want to illustrate this by looking at two near-future dystopian plays,[1] namely Alistair McDowall's *X* and Zinnie Harris's *How to Hold Your Breath*. Each drama focuses on a specific aspect of anxiety: McDowall's on the flowing nature of time and Harris's on its painful, multi-layered and overloaded nature. Both plays also stage anxiety through the crossing of boundaries on a physical and mental level as well as aesthetically through the destruction of language and the ensuing semantic void. Thus, they illustrate that the key to portraying fear and anxiety in dystopian theatre, as a means of commenting on global crises and catastrophes, lies in the crossing of several borders in time and pain and in the destruction of language.

---

[1] For a thorough analysis why both dramas can be characterised as near-future dystopian plays see Trish Reid (79–82).

https://doi.org/10.1515/9783110758252-013

Pain and the Temporality of Anxiety

Fear and anxiety are two closely connected phenomena which can be characterised by similar underlying structures connecting time, pain and language and can function as commentary on global crises in drama and performance. According to Sara Ahmed, the connection between fear and anxiety takes place on a phenomenological level. Similar to Kierkegaard (41–42) and Heidegger (180–181), Ahmed defines fear and anxiety in relation to an object. As she argues, in fear, the object that is feared might pass by, which means that feelings of fear cannot be contained: "If fear had an object, then fear could be contained by the object. When the object of fear threatens to pass by, then fear can no longer be contained by an object" (65). The object of fear is thus *"not quite present"* (65). Therefore, Ahmed characterizes fear as the temporal *"'passing by'"* (65) of a fearful object over time. Even when the fearful object passes by, for instance when a fright-inducing snake vanishes, the fear lingers on, ready to emerge again whenever a snake is seen. Anxiety, on the other hand, can be described as a conglomeration of several of these objects or thoughts that induce fear (65). Thus, it is not a lack of objects but their overabundance which causes anxiety. This can be intensified when each object of fear is substituted by another object of fear over time, which leads to an even bigger conglomeration of fearful objects and ultimately to anxiety. In anxiety, the attachment to these countless objects of fear is so strong that one loses track of what exactly it is one is afraid of. As Ahmed writes, "we could consider how anxiety becomes attached to particular objects, which come to life not as the cause of anxiety, but as an effect of its travels" (66). In the theatre, this cluster of objects and anxiety finds its aesthetic counterpart in what may be termed the multimediality of presentation itself (Pfister 6), where lighting, costumes, props, actors, sounds and speech are all happening at once and acting upon the audience. On a thematic level, this conglomerate surfaces in the various topics and actions that happen simultaneously, and often cross the borders of what can be endured by the characters in a play.

When one follows Ahmed's definition of anxiety as a gathering of several objects of fear, the staging of pain and time must be key to representations of anxiety on the stage. After all, for her, a fearful object that may also be the cause for anxiety is "an anticipated pain in the future" (65). While fear and all its side effects like paralysis or palpitations are felt in the present, the anticipated pain is felt in the future. Therefore, notions of pain and time are essential for analysing anxiety in dystopian plays.

Pain plays a central role in human experience. According to Ahmed, feelings of pain are important for the process of determining the boundaries of bodies;

> [i]t is through sensual experiences such as pain that we come to have a sense of our skin as bodily surface [...], as something that keeps us apart from others, and as something that "mediates" the relationship between internal or external, or inside and outside (24).

While here pain is seen as an indicator for the transgression of such bodily boundaries, Ahmed also establishes the connection between pain and feelings like fear, anxiety and uncertainty when she refers to fear as "an anticipated *pain in the future*" (65, emphasis added). This interconnection becomes particularly forceful, whenever pain is consciously reflected on: "The experience and indeed recognition of pain *as pain* involves complex forms of association between sensations and other kinds of 'feeling states'" (23). Therefore, in pain the borders of body and mind are inadvertently and forcefully crossed on several levels, as the analysis of both dystopian plays further demonstrates.

One important aspect of pain is the inexpressibility of physical and mental pain and the subsequent political consequences this has, as described by Elaine Scarry. Pain, for her, combines the affirmation of our own physical nature with uncertainty about the physical nature of others: "To have pain is to have *certainty*; to hear about pain is to have *doubt* (13). Furthermore, it is objectless and thus "has no referential content" (4). This becomes especially clear when looking at language or rather the inexpressibility of pain through language: "Physical pain does not simply resist language but actively destroys it, bringing about an immediate reversion to a state anterior to language, to the sounds and cries a human being makes before language is learned" (4). This inexpressibility therefore not just influences the experience of pain within someone's personal story: "the relative ease or difficulty with which any given phenomenon can be *verbally represented* also influences the ease or difficulty with which that phenomenon comes to be *politically represented*" (12). Painful experiences, or experiences that induce anxiety, thus frequently remain outside political discourse. The theatre, on the other hand, is a platform where feelings of pain and anxiety can be negotiated due to its status as an aestheticised art form. In the dramatic text, language can actively be destroyed to demonstrate the inexpressibility not just of pain but also of fear and anxiety on an aesthetic level and in wider discursive contexts. Dystopian drama is especially fruitful for this, not only because it is intrinsically related to aversive emotions like fear and anxiety, but also because "dystopian literature generally also constitutes a critique of existing social conditions or political systems" (Booker 3).

What is more, the near-future dystopian scenarios prevalent in plays connected to anxiety also underscore the temporal aspect of fear and anxiety, reflecting Ahmed's notion of a fearful object as "an anticipated pain in the *future*" (65, emphasis added). Bernhard Waldenfels extends this concept of anxiety and time. For him, anxiety combines past, present and future.[2] The object that is acting upon us is too early, for which we are not prepared. Thus, we do not have an answer (and hence no protective mechanism) for an event that suddenly comes upon us. We are afraid that an unforeseen future that changes our way of life might happen too early and become our past without us having an adequate solution for it, ideally to prevent it (Waldenfels 24, 83). While being in the present, we are then at the same time thrown into the future (an event is approaching too fast) and into the past (we are not ready for the unforeseen event that passes us by).

Similarly, in theatre, past, present and future are intermingled, and the linear time structure is broken. As Simon Critchley observes,

> [i]n tragedy, time is out of joint and the linear conception of time as a teleological flow from the past to the future is thrown into reverse. The past is not past, the future folds back upon itself and the present is shot through with fluxions of past and future that destabilize it (32).

This likewise holds true for the near-future dystopian plays that will be discussed below. One way of expressing the dissolution of time in drama is, again, through the destruction of language. Thus, anxiety, as a commentary on global crises and as a conglomerate of fearful objects, can be conveyed in drama and performance through pain and the crossing of various boundaries which it induces, as well as through the connection of future, present and past. In both cases, the sensational and the temporal, this crossing is accompanied by the destruction of language and the depiction of the unsaid through aesthetic means. This is the case in both *X* and *How to Hold Your Breath:* both plays convey their political message through an atmosphere of anxiety, created by the demolition of temporal and linguistic orientation and a pervading sense of pain.

---

2 A similar account can already be found in Edmund Husserl's *Vorlesungen zur Phänomenologie des inneren Zeitbewusstseins*, where he insinuates that perception itself is always in relation to retention and aspiration and thus combines past, present and future (385). Within anxiety, the linear structure of past, present and future is disrupted; as Waldenfels outlines, in anxiety the future has become the past (83).

X: Temporal Disorientation and the Breakdown of Communication

Alistair McDowall's two-act play *X* (2016) is set on a small research base on Pluto, and while crew members Clark, Cole, Ray and Gilda should have returned to earth weeks ago, all technical devices for time measurement and communication are dysfunctional. Cut off from human civilisation, the main characters are driven to conspiracies, anxieties and insanity. In McDowall's "sci-fi horror" (Sierz) both time and language are demolished and simultaneously connected on several thematic and aesthetic levels, thus breaking numerous boundaries and mirroring the multi-layered structure of anxiety. As the title suggests, the letter X plays a central role within this play. It is simultaneously used as a symbol for hugs and kisses, as a placeholder for names and the identity of the characters (McDowall 120), as a visual image smeared across a wall (10) or generally space (based on physical equations). Moreover, as Trish Reid outlines, "[i]t stands for the chromosomal inheritance a mother passes to her daughter [and] is a harbinger of doom in the vision of a little girl someone sees at the porthole" (82). In the following discussion, the focus will be on X as a placeholder for time and a symbol for the destruction of language.

The perception of time within the play is disrupted on several occasions. From the beginning onwards, time is perceived differently by each character and increasingly desynchronised from the time displayed by the large digital clock on the wall. The first scene not only sets the mood for the whole play but is an example of how, in anxiety, time and language are connected and destructed simultaneously. It becomes clear that for unknown reasons the characters have no way of leaving the research base or getting into contact with earth. There is no solution for their technical problems. The characters' involuntary isolation in combination with their different perceptions of time makes them unable to work together. Their relationships are increasingly built on suspicion and resentment, social boundaries are not respected, and, due to the lack of trust, the crew members feel inhibited about showing any emotions (8). Their anxiety steadily increases as they are constantly waiting without any aim or hope:

> Waiting for someone to pick up the phone. Or come get us. ... Or we're just waiting to die. [...] And there's nothing you can do about that. We've got more than enough food. Water won't run out. And the base is designed to last for *decades*, it'll still be breathing *way* after we've stopped. Its *job* is to live forever. (45)

The temporal contrast between the technical devices that last forever and the finite nature of humans can be seen as another source of anxiety in which not just nature but also technology might outlive humans, thus adding to the opaque conglomerate of fears and anxieties.

However, while the technological devices last forever, they may still be dysfunctional. When Cole is doing some math calculations with pen and paper, Clark looks at them and tells him that they are wrong. When Cole explains that he wants to calculate "time. X is time" (66), it becomes evident that all clocks on the base are dysfunctional: "Everything's linked to Earth through the main clock. And the main clock's wrong" (68). Their time neither fits earth time nor plutonian time, which is why Cole's calculations can never be correct. Due to these irregularities, there are constant shifts between day and night, week and month and they do not seem to follow any pattern. Thus, the whole crew does not know what time or day it is, which further increases their uncertain and anxious mood. This uncertainty fits in well with Ahmed's description of the object in fear as "*not quite present*" (65) and to anxiety as an accumulation of these fearful objects which build up "until [anxiety] overwhelms other possible affective relations to the world" (66). While the characters in the play are provided for by the base, they are surrounded by anxiety and fear a future that promises to be an infinite regress of boredom, despair and isolation. Their anxiety, elicited by their isolation and temporal displacement, leads to anti-social behaviour that further isolates them; the characters are not able to speak about their emotions, which still heightens their anxiety.

The only measurement of time in *X* is through dialogues and memory – something that is also increasingly disrupted. Although the drama is set in the future, the characters are constantly trying to hold on to a long gone past. They play games like *Guess Who* from the 90s and chess (28), tell each other stories of the last time they ate meat, of the last birds and trees and listen to the recording of bird songs to remember them and their Latin names (25). This nostalgia contrasts with the futuristic and sterile environment they are in. In an earlier scene, when Clark tells the story of the last tree on earth, he insists that "[h]istory is bullshit" and that one "[c]an't see it. Touch it. There's just this second, right now, as I'm saying it it's dying, it's gone. [...] Pimps like me live in the present" (15). Nevertheless, he is then fascinated with birds, their songs and names (25). This shows that he is torn between a longing for an irrevocable past, the construction of a quasi-pastoral history and an at least superficial rejection of the intangibility of history, an inner turmoil that explains his inability to talk about emotions and sentiments. When he is engaging with birds, language and sounds help to remember a long-gone time and give him some stability and hope. However, this hope is set in the past. Following Waldenfels, the catastrophe that happened prior to the first act and that explains the depletion and destruction of nature on earth, came too soon for humanity to react and prevent it. The research base on Pluto is only the futile attempt to fix the mistakes made on earth centuries earlier. It becomes clear that the crew members have taken their own (and earth's) problems with them. Correspondingly, Michael Billington observes that "the human race, having wrecked its own planet, now transfers its problems to the colonised outer reaches of the solar system" ("X"). Therefore, in *X*, the shifting boundaries of past and future and the inability to live in the present comment on the current political inability to react to several global crises at once.

From act two onwards the division into acts and scenes breaks down. There are no more numbers for new scenes, which are now indicated by brackets [], mirroring the research base trapping its own crew members and the vacuum, isolation and nothingness inside the space station as well as the emotional emptiness inside the crew members' heads. The characters drift apart more and more, and so does their language. Communication in the play is increasingly disturbed as the crew members still cannot contact planet Earth nor communicate with each other, which impacts on their mental states. "In the hallucinatory second act," as Reid writes, "X represents the crossing out of neurons in a dying brain as it colonises language itself, erasing meaning as it goes" (82). Everything becomes very absurd, fast, circling around the same lines of conversation. Ray has tragically died, and Cole and Gilda are constantly in dispute, while Cole has advanced cancer which impairs his mental and physical abilities. Here a spiral begins of Cole forgetting about his illness, growing hostile towards Gilda and later Clark, with his ever-worsening physical and mental state eventually leading to his death. The words and actions crash into each other, breaking the boundaries of mental and physical states, of true and false, illness and health, conscious and unconscious, friendship and hostility. In these scenes the boundaries of time, language, individuality and humanity are breaking down (110). Fuelled by anxiety, the words get mixed up, blurred, and language and time stop making sense. Suddenly feelings are expressed instead of a message, as things are too horrendous to articulate. In the end the crew members cannot remember their old stories, their own names, or what X stands for. The blurring of sensory experiences becomes increasingly clear in the sequence of pages[3] which consist of hundreds of X's, like a wall or a prison of letters that cannot be overcome, a system that cannot be used and experiences and emotions that cannot be described

---

[3] This part of the play contains no stage directions or dialogue markers and thus feels as if it was lifted from an experimental novel, shifting between different genres and writing styles and leaving it open how this is to be staged.

(126–130). On these pages time, language, space, sensation and emotion are all intermingled and cannot be differentiated or expressed, leading to a conglomerate of fear and anxiety.

In the final scene of Act Two, scene X, Gilda is playing hide and seek with a little girl (142). The girl, who was earlier muted with an X across her mouth, is called Mattie. She is not just Gilda and Clark's child but also a hallucination that accompanies Gilda throughout the play. Mattie is thus the connection between Gilda's imagined fifth crew member and her daughter, whose age is shifting between a young girl and a grown woman. In this scene, which probably takes place long after the other crew members are gone, scientific calculations of time and space, age and height once again play a central role and, again, cannot be calculated. The entire conversation between Gilda and Mattie unfolds in temporal loops and circles around the ever-same topics and sentences that constantly shift in meaning. It seems Gilda is as confused by the different time scales as the audience. However, what remains is the love between child and mother. Mattie brings pillows and blankets to put in front of the window for her mother, and it seems as if Gilda mistakes her child for her mother and at the same time Mattie cares for her mother as for a child. The whole scene is riddled with opacity:

> **Mattie** [...] How's that? Warm enough?
> **Gilda** Are we still waiting?
> **Mattie** No, we're not waiting for anything.
> **Gilda** We're not?
> **Mattie** No.
> **Gilda** Oh.
> ...
> *Pause.*
> I don't know what time it is even.
> ...
> And I can't see anything out this window. (148–149)

This scene, once again, shows how time connected with an anticipated painful future is closely linked to feelings of fear and anxiety. Even as an old woman, Gilda is still waiting for something bad to happen. This illustrates Ahmed's claim that the object in anxiety can never be contained because it passes by (65). Gilda is so obsessed with a future of isolation and separation that these feelings have long taken over her life. Therefore, she is not just unable to see anything out of the window, as everything is steeped in blackness; she is also surrounded by an uncertain and already past future that leads to a life of anxiety. As Marissia Fragkou observes, in *X* "the unimaginable or the dystopian serves as an index of what we are about to lose" (91). Alistair McDowall's *X* can thus be seen as an allegory for climate change, for gender relations, for fear of the future, for social isolation and for fake news. In this play several (temporal) boundaries are crossed, which is accompanied by the dissolution of language; the boundary between emotion and reason, nature, culture and technology, old and young, mother and child, noise and silence, innocence and guilt, fear and anxiety vanishes.

How to Hold Your Breath: Social Collapse and Ubiquitous Pain

Zinnie Harris's dystopian alternative reality *How to Hold Your Breath* (2015) uses a similar crossing of boundaries, even though here my focus is on pain and language in combination with anxiety. The play describes the flight and ruin of Dana and her pregnant sister Jasmine from an economically, politically and ethically collapsing Europe: a conglomerate of dreadful events and tragedies happen at once, crash into each other and cross several thematic and aesthetic borders, while feelings of anxiety and pain are increased by and simultaneously cause the destruction of communication and language.

The first border that is crossed in the play is the one between pleasure and pain. At the beginning of the play Dana is in her bedroom with Jarron, who works for the United Nations. It becomes evident that the two of them had a one-night stand. However, Jarron was under the impression that Dana is a prostitute. When he wants to pay € 45 for the night, for her "unlocking services" (50) and Dana refuses, he gets angry. Infuriated that their encounter was not a commercial act but took place out of affection, he insists that he is "unloveable," "a demon" and "a bridge that you don't cross" (23). Dana still rejects the money but is subsequently obsessed by Jarron and persuaded that he is involved in all misfortune that is about to happen to her. The first scene thus gives an impression of the physical and mental borders that are crossed throughout the play – through sex, manipulation and commercial interests. Dana's role as a protagonist is to resist corruption, which, together with an overarching feeling of anxiety, leads to her and her sister's mental and physical annihilation.

After the encounter with Jarron, Dana leaves for a job interview. During the interview she sits on a chair in the middle of the stage, while a bright light is blinding her so that she cannot see any of the interview panellists, who consist of dehumanised voices that surround her (35). This first interview situation establishes borders by playing with darkness and light. Throughout the play, these borders will be demolished as the anxiety-ridden situation of the interview is presented in intermittently recurring scenes. Dana is separated from any human interaction, and the blinding lights that seem to shine through her distract her from concentrating on her presentation. Nonetheless, she is invited to yet another interview to Alexandria.

When Dana and Jasmine, who accompanies her, are on the train to Alexandria, the ticket inspector tells them that the bank has refused their cards. In the meantime, Europe has experienced a major financial and economic crash, leading the surrounding countries to close their borders. Due to their lack of cash, Dana and Jasmine have to leave the train at a place aptly called Hartenharten. In their hotel room, Dana finds out that all the banks have shut. Their hotel room is freezing, the heating does not work, and neither does the kettle. They have no way of getting money or paying for the room. A succession of tragic events unfolds until in Scene Seventeen they have no money, no water or food, no clean clothes, no health insurance and no opportunity to travel to Alexandria. At the same time, they have lost their phones and thus cannot contact anyone, they lost their suitcase with personal belongings, and Jasmine lost the baby, which means that the fear and pain Dana and her sister have to endure occur on both a physical *and* a mental level. There are so many instances of fear and pain that Dana is not able to concentrate on any of them, feeling overwhelmed by what in Ahmed's terms may be described as the conglomerate of fear surrounding her and losing track of what she is afraid of. As Julia Boll claims, "[o]n a structural level, *How to Hold Your Breath* points at multiple causal entanglements that not only affect the characters' choices and their trajectories, but also make up the core structure of the society and indeed the universe in which they move and operate" (234). These "multiple causal entanglements" are, however, obscure. "[B]y far the most awesome and fearsome dangers," as Zygmunt Bauman observes, "are precisely those that are *impossible,* or excruciatingly *difficult,* to anticipate: the unpredicted, and in all likelihood *unpredictable* ones" (11). This further explains Dana's desperate situation and her fight for her own and her sister's life by adding a layer of insecurity, vulnerability as well as chance and unpredictability to her already anxiety-struck situation.

While Dana is surrounded by closing state borders that entrap her during her journey, her individual borders are further invaded. Desperate for her life, she sees prostitution as the only way out of her and her sister's predicament. In the end, her "customer" is not willing to pay the € 45 she demands and rapes her for € 10. While Dana is having rough intercourse with the punter, she is simultaneously asked to repeat her presentation for the interview panel. Being in the spotlight, surrounded by dehumanised voices that have no mercy on her while her body is being intruded is the ultimate crossing of borders in body and mind which induces long lasting pain as well as desperation and anxiety for the future. Pain and anxiety, which, following Ahmed, is a conglomerate of anticipated pains in the future, are transgressive phenomena because they "involve the violation or transgression of the border between inside and outside" (65). In Harris's play, these feelings are accompanied by the dysfunction and destruction of language and communication.

The annihilation of semantic meaning and language can already be traced in the prologue,[4] in which Dana "speaks to the audience" (13), describing the journey she is about to begin. The entire prologue reads like a poem, and the density of the poetic language makes the anger but also the fear and anxiety become palpable. This is the first incidence in which language slips from Dana's lips, and she claims she is "a scream. A howl" (13), comparing herself to wild animals and natural elements. In Harris's play, the dysfunction and destruction of language is depicted in several instances on numerous levels, for instance by Martha and Clare, who can only communicate through violence and beat up an already hurt Dana after the encounter with the punter, or by Jasmine, who, after having lost her child, suffers from a form of amnesia and cannot follow any communication. However, in the following analysis, the focus will be on the interaction between a librarian and Dana. Scene Five is the first time Dana contacts the librarian, asking for a book on daemons after having had the encounter with Jarron. This first interaction takes excruciatingly long as the librarian does not seem to understand Dana's simple request and, by asking unhelpful questions, complicates the matter without offering real help. While this seems to increase the comic relief in the play, the character's role changes in the rest of the play.

In Hartenharten, at his weekend job, the very friendly but pedantic and at times importunate librarian exclaims that he has the books Dana ordered and added a few new ones – alluding to her financial situation without Dana realising what is happening around her. The librarian not just comments here but also foreshadows the events of the play. In a similarly helpless situation, he insinuates that Dana's sister is in despair but does not give any further hints that she is having a miscarriage (105–106). This mixture of foreshadowing and commenting on events in the play comes close to the role of an omniscient narrator in a novel. He is part of the plot and at the same time not involved in what is happening. For Dana, the librarian is the element that crosses the boundary between reality and fiction, someone who is on her side but, although he is trying to help, is utterly useless and counterproductive, thus further adding to her pain. The librarian's comments on the play reach a sad climax when Dana decides to prostitute herself and asks the librarian for advice, who immediately produces "how to" books: "*How to Stop Gagging with Someone's Putrid Penis in Your Mouth* [...]. *How to Make Sure You Don't Get Strangled. How to Not Get a Disease that Will*

---

4 The Prologue was not included in the first production of *How to Hold Your Breath* (Boll 218).

Kill You. How to Stay Alive during Prostitution" (129). Thus, the horrendous events that happen to Dana are imagined, commented on, contextualised and made vivid by book titles. They not only "symbolise [...] a consumerist belief in easy solutions to every problem" (Billington, "How to Hold Your Breath") but also add a visual dimension to Dana's anxiety and pain, commenting on her suffering on a meta-level which further explains the horrible situation she is in and simultaneously prevents any real conversation about it.

Another example of the dysfunctional communication in the play occurs when Dana calls an ambulance for the bleeding Jasmine, who is having a miscarriage. The woman from the emergency hotline does not see the urgency of Dana's call and goes through a slow and inefficient assessment with her, while the librarian is providing Dana with well-meaning but utterly useless advice. Finally, at the end of the assessment, the woman concludes:

Woman
twenty minutes now
Dana
twenty minutes, there must be a way to call a devil
Librarian
you can't rush me
how to keep your cool when your sister is dying
Dana
she isn't dying
can you get it any quicker than twenty minutes?
Librarian
she doesn't get blood, she'll die
Dana
They'll give her blood then, won't they? Someone give her some blood. Give me a book, how to make them give someone blood when they need blood
Librarian
how to listen when people are talking nonsense
Dana
what sort of nonsense?
Woman
I am sorry I have to ask, do you have insurance?
Dana
insurance?
this is an emergency –
Librarian
how to keep your cool when life is stressful
Dana
– I don't need insurance
Woman
do you intend to pay for her treatment in cash?

Librarian
It's got a CD, this one with breathing exercises
meditation (112–113)

In this scene, language is reduced to absurdity, and while Dana becomes increasingly desperate, she does not seem to be able to communicate on a rational level with anyone. Her anxiety and pain thus become evident in the destruction of language in the play. At the same time the boundaries between the reality of the woman from the hotline and the librarian's fictionality are crossed, creating a multitude of unconnected sentences and several threads of communication happening at the same time. This chaos of topics and simultaneous events leads to what Ahmed describes as a conglomerate of different fearful thoughts that merge into anxiety.

At the end of the play, Dana is on a lifeboat with her sister and hundreds of other people, discussing their future perspectives:

Jasmine
we won't ever be going back home, will we?
Dana
I don't think there is anything left for us there.
Beat.
Jasmine
I don't like the idea of not existing.
of being a person but not a person. Like the baby
Dana
the baby –
Jasmine
is dead I know, whereas we –
we'll just be illegal. I understand.
Dana
when we get there, it will get better.
it will all feel better
There is a sudden jolt. (148)

When the boat capsizes, Dana is drowning, while simultaneously finding herself back in an interview scene. She is holding her breath and at the same time trying to answer the questions from the panellists. Talking will drown her and holding her breath will diminish her chances of a new life in Alexandria. This is the most dramatic way of destroying Dana's language. As Elaine Aston observes, "[t]he circularity of Dana's journey from ignorance, through the seeing and to non-seeing augers [sic] a cycle that needs to break, but is not broken." (305) What Dana has to endure throughout the play, her pain and inability to express her needs and anxieties comes very close to what Bauman describes as the three kinds of dangers that induce fear and anxiety:

> Some threaten the body and the possessions. Some others are of a more general nature, threatening the durability and reliability of the social order on which security of livelihood (income, employment), or survival in the case of invalidity or old age depend. Then there are dangers that threaten one's place in the world – a position in the social hierarchy, identity [...] and more generally an immunity to social degradation and exclusion. (3–4)

At the end of the play, Dana has encountered all three kinds of fears and anxieties, as she neither possesses any valuables nor her own body, which has been raped and beaten up, she has no employment or stable environment that will secure her future, and she has no position in society, being a refugee in a foreign country that does not welcome her.[5] Thus, while the destruction of society is described on several thematic levels that crash into each other, thereby commenting on dysfunctional societies, Dana has to endure the crossing of mental and physical boundaries and the destruction of her language, subsequently leading to and being a sign of her pain, fear and anxiety. At this point she really represents "the abyss into which people dread to fall" (13).

Therefore, in Harris's dystopian alternative reality, the political, economic and ethical destruction of society is shown on a thematic level, while these events happen on several dimensions simultaneously and crash into each other, invading personal and social boundaries and inducing pain and anxiety. Similar mechanisms are at play in McDowall's *X*, where the technical devices for communication and time measurement are dysfunctional, leaving the main characters in isolation, desperation, insanity and anxiety. Aesthetically, this is depicted by the simultaneous dissolution of time and language, connecting past and future and breaking several boundaries in the process. Thus, both plays illustrate how fear and anxiety can be represented in dystopian plays through the crossing of several borders in pain and time as well as through the destruction of language and communication.

---

5 Among the manifold anxiety-inducing crises modern society is facing, the crisis of capitalism, whose power to shape our society is increasingly coming under scrutiny, is very prominent. It is also present in *How to Hold Your Breath*, where the banking system fails, and all consumerism becomes dysfunctional. Dan Rebellato has convincingly argued that British theatre adopted a new "anti-realist apocalyptic tone" (para. 4) as "counter-strategy to capitalist realism" (para. 58). For him, this trend is accompanied by a new, less realist representation of violence which is, however, bigger in scale, and as such verging on the apocalyptic.

Works Cited

Ahmed, Sara. *The Cultural Politics of Emotion*. Edinburgh: Edinburgh UP, 2014.
Aston, Elaine. "Moving Women Centre Stage: Structures of Feminist-Tragic Feeling." *Journal of Contemporary Drama in English* 5.2 (2017): 292–310.
Bauman, Zygmunt. *Liquid Fear*. Cambridge: Polity, 2013.
Billington, Michael. "How to Hold Your Breath Review – Magnetic Maxine Peake Is Bedevilled in Morality Play." *The Guardian*, 11 Feb. 2015. Web. 14 Oct. 2021. <https://www.theguardian.com/stage/2015/feb/11/how-to-hold-your-breath-review-maxine-peake-royal-court-theatre>.
Billington, Michael. "X Review – Pressure Builds on Crew Marooned in Space." *The Guardian*, 6 Apr. 2016. Web. 16 Feb. 2021. <https://www.theguardian.com/stage/2016/apr/06/x-review-royal-court-london-alistair-mcdowall>.
Boll, Julia. "Entanglements: Transaction and Intra-Action with the Devil in How to Hold Your Breath." *Affects in 21st-Century British Theatre: Exploring Feeling on Page and Stage*. Ed. Mireia Aragay, Cristina Delgado-García, and Martin Middeke. Cham: Palgrave, 2021. 217–237.
Booker, Marvin Keith. *Dystopian Literature: A Theory and Research Guide*. Westport, Conn.: Greenwood, 1994.
Critchley, Simon. "Tragedy's Philosophy." *Performing Antagonism: Theatre, Performance & Radical Democracy*. Ed. Tony Fisher and Eve Katsouraki. London: Palgrave, 2017.
Fragkou, Marissia. *Ecologies of Precarity in Twenty-First Century Theatre: Politics, Affect, Responsibility*. London: Bloomsbury, 2019.
Harris, Zinnie. *How to Hold Your Breath*. London: Faber, 2015.
Heidegger, Martin. *Being and Time*. Trans. Joan Stambaugh. Ed. Dennis J. Schmidt. Albany, NY: State U of New York P, 2010.
Husserl, Edmund. *Vorlesungen Zur Phänomenologie Des Inneren Zeitbewusstseins*. Ed. Martin Heidegger. Halle, Saale: Max Niemeyer, 1928.
Kierkegaard, Søren. *The Concept of Anxiety: A Simple Psychologically Orienting Deliberation on the Dogmatic Issue of Hereditary Sin*. Trans. and ed. Reidar Thomte. Princeton, NJ: Princeton UP, 1980.
McDowall, Alistair. *X*. London: Bloomsbury, 2016.
Pfister, Manfred. *The Theory and Analysis of Drama*. Cambridge: Cambridge UP, 2000.
Rebellato, Dan. "Of an Apocalyptic Tone Recently Adopted in Theatre: British Drama, Violence and Writing." *Silages Critiques* 22.1. (2017). Web. 16 Feb. 2021. <https://doi.org/10.4000/sillagescritiques.4798>.
Reid, Trish. "The Dystopian Near-Future in Contemporary British Drama." *Journal of Contemporary Drama in English* 7.1 (2019): 72–88.
Scarry, Elaine. *The Body in Pain: The Making and Unmaking of the World*. New York, NY: Oxford UP, 1987.
Sierz, Aleks. "X. Royal Court." *Aleks Sierz*. 20 Apr. 2016. Web. 25 Feb. 2021. <https://www.sierz.co.uk/reviews/x-royal-court/>.
Waldenfels, Bernhard. *Sozialität und Alterität: Modi sozialer Erfahrung*. Berlin: Suhrkamp, 2015.

Peter Paul Schnierer
Visions of Hell in Contemporary British Drama

Hell is the original dystopia, and like all places of superhuman wretchedness it is badly mapped. Knowledge of hell comes to us not via the travel logs of explorers (although Dante sometimes gets close), nor through more systematic surveys but in visions and revelations.

Visions of hell – at first that sounds more determined than the well-worn phrase warrants; many of us will readily imagine an extremely hot or cold place of punishment for human sinners presided over by some manifestation of evil. Yet the iconography of hell in Western culture is too facetted, too complex and sometimes too contradictory to allow of a narrow image of Hell with a capital H, and its topography comprises Tartaros and the Inferno, Sheol and Gehenna. In post-classical literary history, the label can be appended to three distinct genres at successive points in time, with a decreasing dependence on "reality," i.e. an extraliterary acceptance of hell's existence.

The three genres are (1) reports of visions experienced by mediaeval monks and mystics. These reports were – or were claimed and believed to be – records of actual encounters and were read as such, not as accounts of nightmares and hallucinations induced by incense, fasting and mortification. A good example of this genre, amounting to an encyclopedia of infernal tropes, is *The Vision of Tundale*, composed in Ireland in the twelfth century:

> All of the men and the women who descended into the swamp were actually made pregnant by the beast. In this condition they waited harshly for the time agreed on for their departure. The offspring they conceived stung them in their entrails like vipers, and so their corpses were miserably churned in the fetid waves of the frozen sea of icy death. And when it was time, so that they were ready, they filled the depths crying with howls; and so they gave birth to serpents (Tundale 169).

Howls in the abyss, bestiality and miscegenation: as 21st-century readers we might, first of all, recognise the material of H. P. Lovecraft's imagination. Such tropes are equally at home in Dante's *Inferno*, but the *Commedia* turns them into a dramatic purview of the human condition. Its very title associates the text with the stage, the *theatrum mundi*. "Hell on earth" transposes an external, if evanescent reality into a topos occasionally employed by Renaissance dramatists.

(2) The *Commedia*, in turn, is foundational for the second genre, epic poems by post-Renaissance authors, engaging with the progress of enlightenment and its dark flipside. Hell is not a place for the punishment of humans any more, but the pandemonium where the devils themselves are either diminished from powerful beings to "complicated monsters" (226), tangled worms, or else adopt an attitude of solitary contrariness.

(3) The third genre, if that is the right word, comprises works by more recent authors showing a bad state of things with the help of a vast and unspecific metaphor. A typical text would be *Heart of Darkness*, where Conrad loses no opportunity to equate the Congo's upper reaches with various hells, and where he portrays Kurtz as one of "the high devils of the land" (70). Critics are especially apt to use this metaphor as a shorthand for unpleasant conditions; an example of particular interest here is furnished by Trish Reid in her excellent article on imminent dystopias in contemporary drama, where, referencing Raymond Williams, she speaks of hell in two of the plays under review here merely as the depiction of thoroughly impoverished living conditions (78–79).

Tundale and Milton and Conrad, and a host of others, took their diabolical scenarios seriously, in prose and poetry, but drama has always kept an ironical distance. In the mysteries and moralities of the very late middle ages, the first post-classical drama we can recognise as our own, the devil is always already a laughable figure, with hell all but disappeared. This has to do with the homiletic imperative of the church: as impresario of those plays, it had to avoid all suspicions of impartiality, of putting the devil on a level with God, and therefore the point of the devil's powerlessness had to be driven home. Thus he was increasingly portrayed as ridiculous. Even Marlowe's Mephostophilis spends three acts with Doctor Faustus playing pranks on peasants and the clergy. When Ben Jonson, in 1616, wrote *The Devil Is an Ass*, he was able to commit a character assassination: he reduced an already faintly ludicrous stage presence to the role of a victim of human autonomy – mankind is not even aware of him and the few followers that have stayed loyal to him anymore. London is hell enough for them. Similarly, depictions of hellmouth, and of hellfire, belonged to the set of theatrical devices available to the medieval playwrights if they wanted to make a larger point while offering spectacle and frisson. For Jacobean playwrights, who were generally rather fond of unsettling stage representations, the infernal scenery and the special effects had become obsolete as well.

The only component that has survived well is a small number of plots associated with inferno and the devil, such as the story of the Faustian pact. There is no end to the number of Faust plays right up to the present. Another surviving plot is the Harrowing of Hell or more broadly speaking, the *katabasis*, the descent into the underworld, and finally there is the psychomachia, the battle between the forces of good and evil for the soul of Everyman.

I am going to point out a few plays that make use of demonic characters, visions of hell and those plots in more or less recognisable forms. "More or less" because the first specimen, while it looks like a prime example at first glance, is actually the most debatable one. Simon Stephens' *Pornography* (2007) does not deal with the supernatural. It is rather a threnody for the 52 victims of the London bomb attacks of 7 July 2005, who are named and given a brief biographical note each at the end of the playscript. It is up to the individual production, however, to place the necrologue, which is headed "[Scene] One," at any point in the performance, just as the order of the other six scenes remains undetermined at the script stage and the speeches are unassigned. There is no speaker prefix for the following address to the audience either, nor is it given a scene number; it may reasonably be considered a prologue and is quoted here in full:

> I am going to keep this short and to the point, because it's all been said before by far more eloquent people than me.
> But our words have no impact upon you, therefore I'm going to talk to you in a language that you understand. Our words are dead until we give them life with our blood.
> *Images of hell.*
> *They are silent.*
> What you need to do is stand well clear of the yellow line.
> *Images of hell.*
> *They are silent.* (375)

This stage direction, which is repeated four more times in the course of the play, seems to be clear enough, but what do you make of it as a director or designer? Whose images? Dante's? Blake's? Francis Ford Coppola's? To be properly hellish they must be recognizable as such, but if they are, they are clichéd already. It is as if Stephens told the reader or production team: Insert your own illustrated nightmares here. This is hell as dead metaphor, i.e. not hell at all, but merely an indexical sign: Here be monsters. But at least it is an acknowledgement that there is, among the historical and biographical facts the play insists on, an unsayable residue of evil. When one of the speakers confesses that "[t]here are images of things that I have seen seared onto the inside of my skull" (432), he or she similarly points to the limits of language: We are made aware of the trauma but cannot know what it really is and does to the speaker. "Our words have no impact on you," yes, but neither do we have the images to supplant them. Stephens fails to convey the horror, and he acknowledges that failure verbally and visually from the start. This is what makes his play a modernist masterpiece.

Moving from vague images to more specific locations, let us briefly turn to Alistair McDowall's 2015 play *Pomona* (see also Sierz and Tönnies). The starting point of the play's action is this: Ollie's sister has disappeared; she is trying to find her. The setting is the generic dystopian urban wasteland, only this time with a twist: In the heart of this Manchester of ring roads, loops and refracted time and action, there is a dystopian place within a dystopian landscape, a patch of concrete guarded by sinister figures:

> CHARLIE: After following the tip from the man in the car, you head to Pomona and sneak past the two guards.
>
> It is barren and empty and overgrown and you see and hear no one but cars passing in the distance.
>
> You walk on cracked concrete until your foot steps on metal and you find a hatch hidden in amongst the tall grass. (104)

This is meant to sound like an old-fashioned computer game from the time before graphics, and the hatch, like Dante's cave, serves as hellmouth in the old tradition:

> And there's these tunnels that used to have all cables or something to do with the old dock stuff, but they're just tunnels now, and apparently if you follow them they lead to like these huge old air-aid shelters. Like these sort of massive caves they dug in the war. (83)

The tunnels, of course, contain unspeakable horrors, and McDowall acknowledges his debt to H. P. Lovecraft. In fact, references to Lovecraft's Cthulhu myth are frequent and visual as well as verbal. The play was well-received, but as far as I know nobody pointed out what seems – to me at least – obvious: that Ollie's descent into the underworld is the *katabasis* of classical epos. The hero's descent usually follows a quest or is the attempt to reclaim an object or person (or, in the case of Christ's Harrowing of Hell, all the captives in Limbo). *Pomona* presents hell not as a metaphor or allegory, but as a real place. (In fact, it is a real place: you can googlemap it.) It is necessary to point out, however, that the play does not trust its own audacity; it too often lapses into the generic minor-prophet-diatribe we know from so many other texts:

> One day I'll come back to this city on fire.
>
> I'll have flames pouring from me.

> And I'll keep walking through the streets in circles until everyone and everything is just ash.
>
> I'll bring the end to everyone. (62)

Such fantasies of apocalyptic self-fashioning can be traced back to Martin Scorsese's *Taxi Driver*, and they are expressions of helplessness in the face of a world that has become an existential threat. This is why the play offers up a conglomerate of references to the monstrous in popular culture, from Lovecraft's unspeakable cosmic evil via *Dungeons & Dragons* to the Nazi-filled cave in *Raiders of the Lost Ark*. The audience are given visions of hell, but at second hand and incoherently – not because the play is flawed but because that is all there is. This is a creaking, postmodern hell, assembled from whatever embodiment or location of evil is available.

The third play under review here operates differently. In the same year as *Pomona*, 2015, Zinnie Harris's new play *How to Hold Your Breath* premiered on the main stage of the Royal Court Theater in London. The critics were divided; Aleks Sierz spoke of "this dizzy satire of contemporary consumer capitalism [...]. If the final denunciation of naivety is keen-edged, the evening finishes with a soothing quality of mercy that is the show's typically generous gift to us. This sure is New Writing Pure – breathtaking!"

Sierz is a scholarly reviewer who pays particular attention to the text of a performance, and he clearly liked what he heard, but there were dissenting views; Lloyd Evans' review, in the *Spectator*, concluded:

> It's like watching a bombastic lozenge on liquorice stalks bellowing twaddle for two hours. The stirrings and yawnings of the audience put me in mind of fog-bound tourists regretting their holiday plans at Terminal Five. What a night. A state-funded rant advertising misanthropy to sleepy fidgets.

What is it about the play that got a seasoned reviewer so worked up? The plot is straightforward enough: A young woman, Dana, wakes up next to a man she had sex with the night before. He wants to pay her, she is offended, he insists, and eventually he claims to be a demon who must not owe money. She throws him out, while he threatens to make her accept within two weeks. So far we only have his word that he is more than just a psychopath, but things begin to go wrong for her and her pregnant sister, ever more disastrously. At the end of their journey into hell, both women and the child are dead, having undergone the most horrible deprivations and humiliations. In the final scene, the demon – now referred to as such by the stage directions as well – argues with the Librarian, a strange figure who keeps popping up with a cartload of self-help books that are not very helpful. While Dana is drowning, he offers her a book with the play's title: How to Hold Your Breath... In the final argument over her body the Librarian wants her to remain dead, the demon wants to resurrect her; the demon wins that contest, too.

These two clearly are the psychomachia's combatting opponents, the *angelus malus* and the *angelus bonus*, struggling over the soul of a harmless everyman or everywoman, but both of them have issues that simultaneously make them more interesting as characters and more traditional in their comic potential. For instance, they both have the urge to explain and brag about themselves:

> I am unloveable, the unloved. Not the sort of person that gets told they are nice. Feared maybe, fucking hated, yes. I am a devil, I told you, a demon, a thunderclap, I am a really fucking powerful person. People cross the road to get out of my way, I am a nightmare, an underpass in the dark, an alleyway, a bridge that you don't cross. (23)

A bit too much boasting, the metaphors a little overdone. There is something not quite right with this demon, a touch of insecurity maybe. His ancestry includes the sub-sublime devils of the mystery plays, who were to be both attractive and feared – in moderation. But the Librarian is not a proper guardian angel either. Quite apart from being shockingly ineffective, he is also fascinated with the other side:

> LIBRARIAN: [...] you know I once fucked a demon
> DANA: excuse me?
> LIBRARIAN: don't think that you are the only person that has taken that road. Uncomfortable and gorgeous all at the same?
> regrettable and delicious. It was many moons ago, but I remember it well. (69)

These two revenants from the morality plays do not face each other on broadly equal terms, with the Good Angel traditionally at a slight advantage. In Harris's play, his opponent has total control over (wo)mankind, and the comic potential referred to above remains just that: a dramatic strategy that the author all but excludes.

In 2008, the playwright Martin McDonagh made a film that is as foul-mouthed as it is profound: *In Bruges*, a sustained meditation on eschatology, hell, heaven and, specifically, purgatory (O'Brien 101). Two Irishmen, professional assassins, one of whom has killed a child by mistake, are sent to Belgium by their English boss, ostensibly to lie low and weather the uproar. To pass the time they do some sight-seeing; the elder man, Ken, enjoys this unhoped-for holiday, the younger, Ray, feels he is punished for an indeterminate time. He has entered a purgatory of his own making, and gradually he becomes obsessed with the idea. In this scene, they visit the Bruges gallery:

> They end up in front of Bosch's 'Last Judgement,' which we see various details from – freakish demons torturing various people in various freakish ways. Ken and Ray take it all in, quietly.
> RAY: I quite like this one. All the rest were rubbish by spastics, but this one's quite good. What's it all about, then?
> KEN: Well, it's the Last Judgement. Judgement Day. Y'know?
> RAY: Oh yeah? What's that, then?
> KEN: Well, it's, y'know, the final day on Earth when mankind will be judged for all the crimes they have committed. And that.
> RAY: Oh. And see who gets into Heaven and who gets into Hell and all that?
> KEN: Yeah.
> RAY: And what's the other place?
> KEN: Purgatory.
> RAY: Purgatory. Purgatory's kind of like the inbetweeny one. 'You weren't *really* shit, but you weren't all that great either.' Like Tottenham. Do you believe in all that stuff, Ken?
> KEN: Tottenham?
> RAY: The Last Judgement and the afterlife and ... guilt and ... sins and ... Hell and ... all that ... ?
> *Ken realises that Ray is really looking for an answer.*
> KEN: Um ... Oh. Um ... (24–25)

Ray is really looking for an answer, and he both gets and gives it eventually. Ken is issued new orders from England; he is to kill Ray because the death of the child was unforgivable. Ken refuses to obey, and in a psychomachic climax sacrifices himself so that Ray can flee. In the final moments of the film, which may be Ray's last minutes, too, he realizes that someone has given his life so that he may live: Ken's calvary may have been the redemption of both of them. The film leaves the ending open, in a more conciliatory way than the plays do. True to its concern with purgation rather than punishment, it offers a way beyond the psychomachia that has to result in winners and losers, and the dystopian forces that offer damnation only.

Purgatory has its site *In Bruges*, just as hell is located in London, Manchester and the Mediterranean Sea. The allegorical openness of all these places appears to be countermanded by their topographical determination. Correspondingly, there is an insistence on authenticity, or possibly a better term, contemporary validity: *Pornography* attempts to come to terms with the scarring events of the London bomb attacks, *Pomona* is a compendium of current popular culture set in a place you can actually map like a computer game (only the surface, though), and *How to Hold Your Breath* could not possibly have been written before the financial crisis and the thousands of deaths in the new Middle Passage. Dana's terrible fate is nothing more than that suffered by refugees in rickety boats. You do not actually need a demon to destroy humans; they do it themselves, as terrorists, torturers or through simple negligence.

It seems to me that, in their very ineffectiveness, those old visions of hell can currently be of service to the theatre: they gesture at our helplessness in the face of newer apocalypses. Better the devil you know. Whether that is a sustainable dramatic strategy or just an exercise in nostalgia remains to be seen.

Works Cited

Aston, Elaine. "Moving Women Centre Stage: Structures of Feminist-Tragic Feeling." *Journal of Contemporary Drama in English* 5.2 (2017): 292–310.
Conrad, Joseph. *Heart of Darkness*. Harmondsworth: Penguin, 1971.
Evans, Lloyd. "*How to Hold Your Breath*, Royal Court, review: Yet More State-Funded Misanthropy." *Spectator* 21 Feb. 2015. Web. 3 Oct. 2021 <https://www.spectator.co.uk/article/how-to-hold-your-breath-royal-court-review-yet-more-state-funded-misanthropy>.
Harris, Zinnie. *How to Hold Your Breath*, London: Faber and Faber, 2015.
McDonagh, Martin. *In Bruges*. London: Faber and Faber, 2008.
McDowall, Alistair. *Pomona*, London: Bloomsbury, 2015.
Milton, John. *Paradise Lost* ed. Scott Elledge. New York: Norton, 1975.
O'Brien, Catherine. "'In Bruges': Heaven or Hell?" *Literature and Theology* 26.1 (2012): 93–105.
Reid, Trish. "The Dystopian Near-Future in Contemporary British Drama." *Journal of Contemporary Drama in English* 7.1 (2019): 72–88.
Sierz, Aleks. "How To Hold Your Breath, Royal Court Theatre." *The Arts Desk* 11 Feb. 2015. Web. 3 Oct. 2021 <https://theartsdesk.com/node/74802/view>.
Sierz, Aleks, and Merle Tönnies. "'Who's Going to Mobilise Darkness and Silence?': The Construction of Dystopian Spaces in Contemporary British Drama." *Journal of Contemporary Drama in English* 9.1 (2021): 20–42.
Stephens, Simon. "Pornography." *The Methuen Drama Book of Twenty-First Century British Plays*. Ed. Aleks Sierz. London: Bloomsbury, 2010. 373–442.
Tundale. "Tundale's Vision." *Visions of Heaven and Hell before Dante*. Ed. Eileen Gardiner. New York: Italica, 1989. 149–195.

Ilka Zänger
"Hiding from the World": Dystopian Subjectivity in Martin Crimp's *In the Republic of Happiness*

"It's distressing. [...] It's deeply serious. [...] It's dark" (Crimp, *Attempts* 205). These words of a nameless voice on an alleged answering machine in Martin Crimp's *Attempts on Her Life* might be well applied to the whole entity of his theatre texts. With the capitalist world of neoliberalism and the all-seeing eye of (social) media looming in the background, Crimp does not wrap his recipients in cotton wool. In Aleks Sierz' words, his work "is characterized by a vision of society as a place of social decline, moral bad faith, and imminent violence" (2). Crimp's bleak visions of society doubtlessly resonate with the idea of dystopia. But how profoundly are they connected with "classical" literary phenotypes of dystopia? How do they deviate? Or do they even exhibit a very specific and unique dystopian appeal? To approach these questions, the following paper will firstly give a very concise definition of classical concepts of dystopia. Secondly, the focus is set on Crimp's theatre text *In the Republic of Happiness* (2012),[1] approaching the dystopian appeal lying at its core. Based on this analysis, the paper will close on a tentative conceptualization of Crimp's dystopian objective.

In Lyman Tower Sargent's concise definition, a dystopian text presents "a non-existent society described in considerable detail and normally located in time and space [...] considerably worse than the society in which [the] reader live[s]" (*Utopianism Revisited* 9).[2] As opposed to utopian writings which normally describe their social *topos* from an outside perspective (Sargent, *Utopianism* 29), the dystopian text begins *in medias res* focusing on a protagonist who "is always already in the world in question, unreflectively immersed in the society, [which is why] cognitive estrangement is at first forestalled by the immediacy and normality of the location" (Baccolini and Moylan 5). The action of the narrative is pro-

---

[1] As Crimp's texts deviate from dramatic convention, they are not referred to as dramas or dramatic texts in this paper. Instead, the term "theatre text" (Poschmann 22) is chosen to indicate their problematization of dramatic convention as well as their orientation towards the theatrical stage – in the broadest possible sense.
[2] Just a few paragraphs above this definition, Sargent gives a more careful wording with regard to the use of detail: He speaks of a society which is described "in some detail," stating that "the completeness will vary" (*Utopianism Revisited* 7).

pelled by trickle-down effects of the dystopian macrocosm on the individual. However, the protagonist develops a critical stance towards the surrounding social system, resulting in "the construction of a narrative of the hegemonic order and a counter-narrative of resistance" (Baccolini and Moylan 5). The classical dystopian narrative is thus constructed around the contrastive pair macrocosm and microcosm, the former being portrayed as a hegemonic social system demanding submission from its citizens who, in turn, consensually participate in its socio-political mechanisms. The latter, microcosm, portrays an individual or a group of individuals in such a way that the narrative directly jumps into action and evolves around the conflicting relationship of the individual with the dystopian society, turning from complicity to rebellion – most likely without general success in overthrowing the system. Generally, the focus on the antagonists, the individual and society, is more or less balanced – the dystopian narrative explicitly relates the underlying mechanisms of the dystopian society, its determination of the people and the negative influence on the individuals justifying their rebellion against the imposed mechanisms.

These basic definitional cornerstones can easily be applied to classical dystopian literature such as Aldous Huxley's *Brave New World* or George Orwell's *1984* – but do they also apply for Martin Crimp's work? Starting the search for the dystopian element in his theatre texts, it is quite illuminating to adopt a broader perspective on the fictional worlds which are evoked. At their very core lies the problematized entanglement of the individual human being with the wider sociopolitical dynamics of neoliberalism, portraying a bleak vision of the world: In *Dealing with Clair*, a young female estate agent mysteriously disappears, yet her death meets with utter indifference. Instead, the audacious accumulation of money from the selling of a mediocre house is what counts. *Cruel and Tender* unfolds against the background of a war, waged by the protagonist Amelia's husband, the General – who has to take responsibility for his crimes against humanity in front of a tribunal. Apparently, he "would massacre a what? – an entire / population?" (29) because he was madly in love with a child. However, he gets off rather lightly as his actions are not severely punished. Generally, the dystopian element is not to be found in the obvious atrocious context of the surrounding society but comes to the surface in individual actions. In *Play with Repeats*, for example, Nick stabs Tony to death on some quite random cause – after having made his acquaintance in a pub just a few minutes earlier. In *Getting Attention*, a child allegedly dies because people turn a blind eye to obvious dangers – the neighbours are on the one hand eager for sensation and spy out everything, but on the other hand do not help when they notice that things are not going well, and the young social worker is fobbed off with all too simple explanations. In *Definitely the Bahamas*, Milly and Frank are portrayed talking about all kinds of riches, travels, cars and money in the life of their son and his wife, while completely ignoring his apparent abuse of the young Dutch Marijke.

The texts show individuals in crises, yet the macrocosmic crisis remains hidden under the surface – nevertheless, the notion of dystopia stays subliminally relevant. In this context, especially *In the Republic of Happiness* catches the eye: The title is, at first, reminiscent of a utopian society – rules and conventions guaranteeing the happiness of people living in this republic. Yet one may sense a certain sarcasm. In fact, already the title of Part 1, "Destruction of the Family," gives the play a clearly dystopian appeal. It is set in a Christmas nightmare in which we follow a family's subliminal interpersonal atrocities that constantly drift towards the absurd in a satirical highlighting of commodity culture. The action starts *in medias res* with a conversation on Debbie's pregnancy – one of the two daughters of Sandra and Tom. When Debbie says, tellingly, "And that's how it should be, Dad – however much I love you, I know that I'll love my baby more. Which is why I'm afraid. Wouldn't you be afraid? When you look at the world? – when you imagine the future?," her sister Hazel indifferently answers: "So why doesn't she just get rid of it?" (277). The topic of Debbie's unborn child conceived from an unknown father is played over by diversionary tactics from Sandra, talking about the supermarket visit and the nice lettuce which the grandmother brought. Hazel, however, breaks the idyll again by identifying another shopping item – the porn magazines for granddad. A few moments later, Hazel devalues her Christmas present – a dress, which "wasn't exactly expensive" – and gets really angry because her sister "made this long long list of all the things she wanted – and because she's pregnant she got them" (281) – among other items, a car and diamond earrings. These expensive presents stand in sharp contrast to the fact that the family is not willing to spend money on electricity: They have unscrewed all the light bulbs, just to insert them again the moment it gets too dark. By this, the absurdity of commodity culture is highlighted satirically: The most basic needs take second place to new, expensive and also pretentious consumer goods which are not really needed. This is further taken to extremes when the grandmother enthusiastically talks about her taxi tours:

> I like to think ah these two minutes in a taxi have already cost me what that man emptying the bins will take more than an hour of his life to earn – and oh the extra stink of a rubbish bin in summer! Yes on nights like that the taxi is glorious and the fact I'm paying for my happiness makes my happiness all the sweeter – and the fact that other people are having to suffer and work just to pay for such basic things as electricity makes it even sweeter still. (282)

Spending money is no longer only a way to the satisfaction of owning something new and beautiful. Instead, the action as such, the sheer ability to spend money, becomes fetishized. This is enhanced by a sadistic enjoyment of the confrontation with poorer people – hypocritically denying the neediness of one's own family. The uncanny atmosphere increases further when Debbie and Hazel give their very special in-yer-face version of a Christmas song which they seem to have made up themselves. They sing:

> We're going to marry a man
> (going to marry a man)
> The man will be rich
> The man will say bitch:
> > I'll make him pay for my meals
> > I'll strut and fuck him in heels –
> That's our incredible plan
> Yes our incredible plan. (286–287)

Material values are of utmost importance and beat an affectionate relationship – which is apparently not reprimanded by the family even in the course of the following stanzas in which the girls sketch killing scenarios and sing about sending their future husband to the moon. Tellingly, this is the very moment Uncle Bob enters the scene to say goodbye to the family: He leaves with his wife to some place unknown, never to return, and just came by to deliver a farewell hate speech from his wife Madeleine, telling the family "how much she hates you" (290). She can't do this herself because "[h]er workload is appalling" (290) – in fact another indicator of the emaciating circle of production and consumption. The climax of Bob's hate speech is the horrible description of Madeleine wanting to bang Sandra's head against the wall until her teeth break in her mouth.

The family's interaction discloses decaying relationships in which the basic rules of human company have been lost. This general deterioration can be best illustrated by the use of language which is no longer the means of conversation on which human connection is built but has turned to an empty vessel of impulsive unfiltered utterances often sounding artificial and bereft of human decencies. The entire conversation seems to be marked by humiliations, devaluations and interpersonal atrocities, showing an underlying lack of empathy. As a result, the characters appear to have lost orientation completely – in a world without profound relations with their human counterparts they no longer know where to turn to. Instead, they hang on to prescribed social stereotypes as patterns of identity with the aim of not getting completely lost. Sandra keeps to prescribed female roles though she does not seem to feel comfortable with them, while the grandmother supports her husband in his assertion that he was the provider of the family although, factually, she herself earned the money. Instead, she plays the stereotypical role of a wife, shopping for groceries and porn magazines for her husband and performing the household chores. Beyond their artificial conversation, the individuals seem to be completely confused and disoriented, almost developing a repulsion for basic human behaviour. Being human becomes abject and undesirable, which is highlighted in the grandmother's words:

> I sometimes wonder if we are on the verge of some enormous and magnificent change – don't you think? Yes I mean a change to our actual human material. [...] Because what I'm imagining – Hazel – in that taxi of mine, is a new kind of magnificent human being who may not even be human at all. (282)

In the universe of Crimp's text, the revolution would be regaining humanity and newly tied emotional bonds to fellow human beings. Yet, attempts to escape play into the hands of materialism: Uneasiness and the lack of well-being are numbed by getting rid of the least bits of humanity. The human being craves becoming material itself. All surface in a beautiful haute couture dress, showing off her wealth and "radiat[ing] charm, charisma, conviction, power" (282), Madeleine states:

> I don't go deep. [...] No. Because this new life of ours – what will it be? [...] Thin [...] as a pane of glass. [...] Hard. Clear. Sharp. Clean. And if any one of you so much as touches it, you'll be cut right through – right through to the bone. (303)

The repulsion for the human basically leaves nothing but an empty void from which people have to distract themselves – by never "going deep," i.e. by turning a blind eye to their innermost human nature.

To fill the void this denial creates, consumption is the need of the hour – strangely menacing in its utter inaccessibility, the consumerist hegemony profoundly infiltrates the human individual, creeping under its skin and forming a pathological conglomeration with it, thus "leading to the ultimate question of whether our lives even belong to us at all" (Angelaki, *Social and Political Theatre* 140). Just as the macrocosmic functional structure merges with the individual human beings, so it becomes clear that an escape from this macrocosm must necessarily remain futile. Serving as a distraction from the profound disorientation which apparently cannot be resolved on earth, space (or a diffusely indescribable yet different "beyond") becomes a leitmotif constantly recurring throughout the text: It appears as a motif in the disturbing song of Hazel and Debbie, the grandfather apparently tried to set up a company for space travels, space rockets are built. Yet salvation is only promised to those who can afford it – in this case Bob and Madeleine. Part 1 ends with Madeleine singing a song with the following last stanza:

> I've booked my ticket: I'm flying first class
> to a cool place thin as a pane of glass
> where I just have to swipe a security pass
> to swim in the milk of thick white stars.
>
> > It's a new kind of world
> > and it doesn't come cheap
> > and you'll only survive
> > if you don't go deep
> > (so I never
> > no I never
> > no I never go deep)
> > (*Republic* 305)

The nightmarish Christmas from Part 1 and Madeleine's and Bob's personal utopia (and at the same time dystopia) of Part 3 are connected by a hinge in which dramatic convention is blown apart. It is written in free form without clearly assigned parts, serving as a metadramatic rite de passage into the Republic of Happiness: Just as they leave behind all the loathsome pains of human everyday life, Part 2, "The Five Essential Freedoms of the Individual," builds what Aleks Sierz describes as a "superlative account of the pains of individualism" (234). The individual's freedoms have in fact become a burden, freedom is nothing positive anymore and turns into a mere *freedoom* interspersed with disorientation and alienation from the world – and this "overemphasis on performing emotion [functions] as substitute for implementing action towards change" (Angelaki, *Social and Political Theatre* 155, with reference to Frank Furedi). In this second part, Crimp elegantly mirrors the nightmarish visions of *In the Republic of Happiness* in the dissolution of conventional dramatic form – it even seems as if the portrayed disorientation and dehumanization has a parasitic influence on the textual structure. Mirroring the characters' lack of orientation, the clear-cut dramatic frame of acts and scenes dissolves into "parts," by which the module-like structure of the theatre text is highlighted. Just as there are no grand narratives and no possible solutions, a clearly developed plot is given up in favour of snippets of action, only loosely tied to each other. And like a comment on Madeleine's fetish for not "going deep," Crimp gives his theatre text the ironic subtitle "entertainment," referring to the shallow anaesthetics of the media culture, or, put another way, the soma of neoliberalist society.

Just as this second part transgresses conventional dramatic form, Madeleine and Bob leave everything behind to enter their Republic of Happiness – a satiric take on "the popular middle-class habit of booking an escape" (Angelaki, *Social and Political Theatre* 145). Part 3, "In the Republic of Happiness," is introduced by a quotation from Dante's *Divina Commedia:* "Tu non se' in terra, sì come tu credi" (a quote from Paradiso, Canto I, 91 – freely translated: You are no longer on earth anymore, as you may still think.). With this introduction, Crimp clearly situates the events in some place unknown to us, a u-topos, promising salvation to those who can afford "hiding from the world" (Crimp, *Attempts* 206) as big players in the cosmos of consumerism. Interestingly, it remains a strangely impalpable, semi-real and floating space, a mere simulation based on illusions that can be shaped according to the inhabitants' desires – yet beyond its surfaces it is uncannily hollow. Madeleine's and Bob's space unveils the pretentious make-believe of the well-functioning family life of Part 1, laying bare what humans have become – individuals out of contact with their innermost selves and their human counterparts and dependent on the outside material world. The description of this place can be illuminated with Michel Foucault's concept of heterotopias, which he describes as "something like counter-sites [which are located] outside of all places" (24). Just as heterotopias maintain a special relation to time, Madeleine and Bob experience an "absolute break with their traditional time" (26). Secondly, "the heterotopic site is not freely accessible like a public space" (26) – in the case of Madeleine's and Bob's escape the means of entering it is large sums of money. It becomes the entrance ticket to an uncanny space which is not the absolutely happy place, but instead in its deprivation of human appeal lays open the deceitful mechanism – the subliminal doctrine of neoliberal society that money can buy happiness. Madeleine's and Bob's personal heterotopia is definitely far away from a vision of paradise: Bob finds himself in an "enormous" and empty room, sterile and impersonal, "[h]ard. Clear. Sharp. Clean," just equipped with an office-type desk.[3] Beyond the large windows, a green landscape is insinuated, yet cannot be seen clearly. And this again ties in with Foucault's concept of heterotopia, which he describes as "hav[ing] a function in relation to all the space that remains" (27). In this case the Republic of Happiness "create[s] a space that is other [...], as perfect, as meticulous, as well arranged as ours is messy, ill constructed, and jumbled" (27). From Bob's conversation with Madeleine we learn that everything in this place works according to her rules, that the two are sovereigns deciding over the fate of their citizens. However, Bob seems to be completely lost – not only physically in the

---

[3] That the office desk is the only piece of furniture can be read as yet another sideswipe at the inflated meaning of money – work, i.e. earning money, finds its way into the private space of the home, so that the exclusion of gainful employment from the private space is dissolved – the human being merges with the role of the productive individual employee in consumerist society.

vast empty room, but also mentally: He suffers from dementia and seems to be stuck in the situation in Part 1, when it was his task to deliver Madeleine's message to the family: "I have so much to remember. [...] The things you've said I'm to say." (347) He seems to be completely dependent on Madelaine as a guide for his existence. She, however, still wearing her haute couture dress, turns from the already cool and detached woman of Part 1 into a mere torturess. What shines through here is Eva Illouz' conceptualization of romantic love as an ideological system, serving and fostering the interests of capitalism (180) – yet while Illouz still wonders if the amalgamation of love and consumption results in a deterioration of the relationship, this is taken to the extremes in Madeleine's cold behaviour towards Bob: She frequently bites him – which even takes on a cannibalistic air, as she draws his blood. She constantly bites him awake when he falls asleep – to find out "if you're happy. [...] I think you've forgotten how happy you really are. I think you are starting to forget how happy this world really makes you – grinding your teeth – grabbing – thrashing. Because what do you want? What have you not got?" (350–351). This seems to confuse Bob even more and we can sense his desperation growing. With an air of complete emptiness and a loss of tactile sensation he says to Madeleine: "You talk about the world but I listen and listen and I still can't hear it. Where has the world gone? What is it we've done? – did we select it and click? – mmm? Have we deleted it by mistake?" (352) A conventional trait of dystopian narratives, the maintenance of illusion by a specific means or drug, is mirrored in the Republic of Happiness, where this function seems to be taken over by the "100% happy song" Bob is forced to sing by Madeleine: When all turns dark in the end he only hears her commanding voice until she appears in a Big-Brother-like cyborgian vision, commanding him to "click on / my smiling face." Bob knows what is expected from him: "Click on my smiling face and you can install a version of this song that has no words at all." (358) Though the ending is left open, the visions are bleak. "In the Republic of Happiness" with its science-fiction-like appeal, human destiny is caught in a process of dissolution inside a Baudrillardian hyperreal: Here, referentials of the so-called real no longer exist but instead have undergone an "artificial resurrection in systems of signs" in which "any distinction between the real and the imaginary" becomes obsolete (Baudrillard 4); and in this hyperreal, which is procured by the constant process of simulation, "truth, reference and objective causes have ceased to exist" (Baudrillard 6). Crimp clearly works with the dystopian motif of the creation of illusions of perfect surroundings, following the innermost desires of people – at the cost of their own humanity and the loss of their real self and presence.

After this concise consideration of *In the Republic of Happiness*, the question remains: How does it connect with a rather classical conceptualisation of dystopia? In Sargent's definition, a characteristic trait of the dystopian narrative is the detailed portrayal of a "non-existent society" (*Utopianism Revisited* 9) which is normally set in worse circumstances than the recipient's. With regard to *In the Republic of Happiness*, however, it becomes difficult to clearly name central characteristics of the society in which Crimp's theatre text is set. As opposed to the typical dystopian narrative, Crimp does not portray it in specific detail; he rather shows the influences this world has on the individuals living in it. In the neoliberal society, the poisonous capitalist greed seeps into the lives of human beings, devouring all those who are too weak. If you are not one of the big players, either owning money or estate, you serve as a mere cog in the wheel and you are passed over in silence. And quite frighteningly, this is a world strangely familiar to us. In fact, Big Brother is missing – the big dark counterpart of the dystopian individual remains hidden under the surface, graspable only in the diffuse mechanisms of capitalism: All seem to seek fulfillment in material things. As Vicky Angelaki points out, "to tie happiness to individualist neoliberalism is a utopia, from which not only isolation, but also dystopias spring" (*Social and Political Theatre* 138): With Stuart Hall's circuit of culture (Barker 22) in mind, we can say that individual identity and subjective agency have entirely merged into the processes of consumption and production. Consequently, instead of portraying the dark macrocosm in concrete detail, the focus is switched to the microcosmic level of the individual subject and its situatedness in an interpersonal social framework which has fallen ill with disorientation. The interaction with human counterparts is completely dysfunctional, lacking the ability to change perspectives and to show empathy. In this context, Nancy L. Nester's considerations of the relationship of empathy to the utopian impulse are enlightening: On the basis of Jeremy Rifkin's quote, "[a] world without empathy is alien to the very notion of what a human being is," Nester refers to anthropologists Richard Leaky and Roger Lewin, who observe that "any factor tending to bind individuals more closely together contributed to the eventual success of the species," and to the biologist Robert Plutchik, who notes that "from an evolutionary point of view, empathy has important survival value" (Nester 117–118). In turn, a lack of empathy goes together with a profound repulsion for human nature, resulting in society falling apart and threatening humanity as such. The human condition becomes the abject; fulfilment is sought beyond human corporeality. In this profound disorientation, the space in which the human is situated is distorted and becomes a mere hollow container, a pure materiality – a locus of self-alienation. The protagonists project their longings to a u-topos situated in a heterotopic beyond – transgressing their own living space.

Another deviation from classical dystopian narratives is the lack of a critical stance towards the morbid macrocosmic structures. In Crimp's theatre texts, we do not find any serious or promising attempts at rebellion. His works portray the moment in which broader macrocosmic structures have already influenced individuals to such an extent that they have fallen ill from this – either in the contact with others or with regard to themselves, suffering from being deprived of humanity. Resistance is no longer an option and in fact becomes impossible. Tom Moylan coins a specific term for this phenomenon: Based on Søren Baggesen's distinction between dystopias exhibiting a "utopian optimism" (63) (i.e. an element of hope/resistance) and those marked by "dystopian pessimism" (63) (i.e. resignation), Moylan describes the latter as "pseudo dystopia[s]" (63). To me, this term is slightly misleading, as it would translate into something like a "pseudo bad place" – and the phenomena which the term is meant to describe basically represent the opposite. Instead, I would opt for the tentative concept "subjective dystopia" to refer either to the shifting focus from macro- to microcosm or to the catastrophe of the individual subject which is highlighted. This newly set focus actually opens up wider angles on the concept of dystopia; Crimp's dystopia is a dystopian crisis of the individual human being.

What is especially noticeable in Crimp's tackling of dystopia is the dissolution of dramatic form in a "radical perceptual defamiliarization" (Angelaki, *Martin Crimp* 12). This, first of all, serves the function of mirroring the theatre text's *fabula*, the "subjective dystopia." The crisis of the subject becomes a crisis of conventional drama. In *In the Republic of Happiness*, this is mainly realized in Part 2, where speech is no longer assigned to clear-cut characters. The sarcasm accompanying the five essential freedoms of the individual (e.g. "3 THE FREEDOM TO EXPERIENCE HORRID TRAUMA" [322] or "5 THE FREEDOM TO LOOK GOOD & LIVE FOREVER" [334]) highlights the illusion of individual freedom in the heteronomous, patronising mechanisms of neoliberal society. Secondly, Crimp's metadramatic technique functions as a means of including the recipients in the negotiation of topical aspects. The suspension of conventional dramatic form leads to a disturbance of illusion, resulting in the audience "seeing double" (Richard Hornby qtd. by Hauthal 52). This shifts the performance more than ever to the here and now and calls for a new degree of reflexivity not only in the process of sense-making but also in the transferal to the audience's personal experiences. In the end, this results in an intensified cognitive and emotional involvement of the spectators (Hauthal 108).

Consequently, with regard to "subjective dystopia," especially those theatre texts seem to be of interest which mirror the crisis of the subject and the illusory character of alleged human agency in the *fabula* as well as on a metadramatic and -theatrical level.⁴ A case in point is *The Rest Will Be Familiar to You from Cinema*. In his adaptation of Euripides's *Phoenician Women*, which evolves around the power struggle of the two brothers Eteocles and Polynices for the reign in Thebes, Crimp changes the ancient Greek choir of Phoenician women into a group of sphinx-like girls – no grown-ups, potentially immature –, which may be a hint at the haphazard and disoriented development of events and at the potential loss of individual agency. And indeed: While in the antique tragedy, the choir generally occupies a characteristic position in-between the protagonists and the audience outside the tragedy and presents songs referring to the dramatic action without ever interfering (Haß 51), the girls entirely hijack the dramatic level of action. The characters' (or better: marionettes') words seem to be completely injected into their minds by the girls – refusal impossible:

> Jocasta: NO! DON'T KEEP TELLING ME WHAT TO SAY!
> I am free. I am free.
> I'm a human being. Look at me. I can say
> what I like.
> *Pause.*
> Girl: Says Jocasta.
> Jocasta: Says Jocasta, wife and mother of Oedipus. (*The Rest* 14)

This stands in sharp contrast to the basic roots of ancient drama where the protagonist is separate from the choir. While this separation may be identified as the birth of "modern subjectivity" (Haß 51), the protagonists in *The Rest Will Be Familiar to You from Cinema* are no longer self-determined. Their actions are not based on their own motives but are prompted by the group of girls.

Next to this metadramatic technique, metatheatrical elements can be found already in the first scene, when one of the girls refers to a specific camera shot of the film *Oedipus* by Pier Paolo Pasolini – which can directly be connected to the title of the theatre text. While this can be considered an ironic and satirical metatheatrical comment, another aspect resonates here as well: *The Rest Will Be Familiar to You from Cinema* is built around a two-fold fictionality in which both dramatic representation as well as the entire course of action is marked by an unreal appeal, leading to the question of reality as such. This is also revealed in the last words of the girls:

---

4 Janine Hauthal differentiates between metadramatic and metatheatrical strategies. While "metadrama" refers to phenomena reflecting the fictional character of dramatic narration without pointing to the media-technological features of dramatic theatre texts, "metatheatre" describes phenomena which highlight the theatrical mediation and textualization in theatre texts (129).

- Where is the world?
- Good question.
- What did you do? Select and click?
- What did you delete? [...]
- [W]as it your own human material? Well?
  Pause.
- What does a Sphinx want plus
  who does a Sphinx fuck when?
  What does a blind man see
  when he looks through the mineral ring of her iris?
- What film do you endlessly project
  in the deserted cinema of my mind? (*The Rest* 84)

This epilogue not only resonates with the last scenes of *In the Republic of Happiness* but also circles around the main underlying question: What is left of the human being in this world? In Greek mythology, the Sphinx, in the siege of Thebes, asks the citizens a riddle, threatening to eat them in case they cannot solve it: What walks on four feet in the morning, two feet at noon, and three feet in the evening? In Greek mythology, Oedipus' right answer to the Sphinx's riddle is: the human being. Yet with the loss of human agency this is no longer a question to be asked; the sphinx-like girls' riddles are unsolvable and absurd.

The notion of the subject is still more complicated in *Attempts on Her Life*. While in *The Rest Will Be Familiar to You from Cinema*, the characters have some fragile performative bodily presence, this is not the case at all in *Attempts on Her Life*. Martin Crimp sketches 17 scenarios here, all dealing with a floating character named Anne – or different versions of this name. While Anne can be described as the protagonist of the theatre text, she never appears in person – she is no "self-identical subject" (Hauthal 250) – and thus, in its essence, leaves a blank space. This blank is filled with completely contrasting identity sketches – all rendered through the eyes of other persons and ranging from social roles such as that of a daughter to a car in a TV-commercial. Anne becomes a mere projection surface (Hauthal 250) for other people's perceptions and transgressive attributions. In a consumerist attitude, she is turned into a vessel for all the desires others want to project onto her. What is likewise at stake is the question of self-representation and the dissolution of self which is central to our culture of medial representation and social media. She is established as a heterogeneous being in a postmodern pastiche of textual surfaces. *Attempts on Her Life* is the identity simulation of an absent character, a heterotopian protagonist so to speak, who is, at the same time, the text's broadest gap. It is a theatre text on the intraceability of agency and thus on the question of the nature of agency, the actuality and reality of individual perception, the impossibility of factuality, the nature of the image as a mediator of reality, and, in the end, the issue of reality as such. In both of these examples, upon which this paper could only touch very briefly, Crimp mirrors the theatre texts' subject matter in his appropriation of dramatic form – and strikingly, both topics as well as the form are scrutinized in the dystopian theatre text *In the Republic of Happiness*. It remains a desideratum to delve more deeply into the notion of dystopian subjectivity by analysing further how the topic of the decay of humanity is mirrored in the specific textual form Crimp chooses for his theatre texts.

The works operate beyond the context of dysfunctional societies which we are all able to infer, exhibiting a subliminal dystopian appeal that might well be described by what Raffaella Baccolini and Tom Moylan – referring to Raymond Williams – call "dystopian structure of feeling" (4). Beyond this dystopian background, the focus shifts towards the level of the individual. Crimp's theatre texts are of uncanny likeness to us, uncomfortably familiar. They make us realise that dystopia has come close to our skin. According to Oliver Nachtwey, we live in a society of social descent, the increasing prevalence of simple and badly paid jobs, the loss of social and solidary relationships as well as the development of precarity – also of highly qualified persons as professional career paths can no longer be planned with dependability (8, 12–13, 110, 121–122, 137, 153). It is a society in which individualisation loses its emancipatory character and instead turns into another challenge or even an imposition – threatening to become pathological as sociality as such is negated. Nachtwey's visions are bleak – and they describe the world in which Crimp's theatre texts are set. These are worlds in which dystopian tropes have become an ultimate immediacy, so that the basic functional mechanisms of dystopian narratives, "speculation and extrapolation" (Voigts 1), seem to move away from the spotlight. It is the world of the here and now, the hegemony of neoliberalism, of commodity culture – a world, from which no escape is possible. Dystopias tell us of societies worse than our own, of the struggle of humans to live – at best to live a good life – against all odds. Dystopias are pessimistic tales – calming us, as we see that there are still ways in which things could get worse. One might ask if these texts function as another opium for the people in times marked by ecological crises, invincible capitalism, and a pandemic which has still not been defeated after more than two years. Yet one might also describe the works as subliminally hopeful tales – helping us uphold the faith in our own power, to believe that there are things yet to be done. According to Vicky Angelaki,

> Crimp's theatre is based on the fact that distinctions between spectatorship and citizenship cannot hold: the two merge in the act of witnessing these plays, which reinvigorate our abil-

ity to acknowledge our being, rights and, equally importantly, responsibilities within our given society, as participation and involvement are key. (*Making Theatre Strange* 12)

Looming above Crimp's texts is the perpetual question: In the world we live in, what is the human being? And no less is at stake: Our humanity.

Works Cited

Angelaki, Vicky. *The Plays of Martin Crimp: Making Theatre Strange*. Basingstoke: Palgrave Macmillan, 2012.

Angelaki, Vicky. *Social and Political Theatre in 21st-Century Britain. Staging Crisis*. London: Bloomsbury, 2017.

Baccolini, Raffaella, and Moylan, Tom. "Introduction: Dystopia and Histories." *Dark Horizons: Science Fiction and the Dystopian Imagination*. Ed. Raffaella Baccolini and Tom Moylan. New York: Routledge, 2003. 1–12.

Barker, Chris. *The Sage Dictionary of Cultural Studies*. London: Sage, 2004.

Baudrillard, Jean. *Simulations*. New York: Semiotext(e), 1983.

Crimp, Martin. *Plays 2: No One Sees the Video, The Misanthrope, Attempts on Her Life, The Country*. London: Faber and Faber, 2005.

Crimp, Martin. *Plays 3: Cruel and Tender, Fewer Emergencies, The City, Play House, Definitely the Bahamas, In the Republic of Happiness*. London: Faber and Faber, 2015.

Crimp, Martin. *The Hamburg Plays: The Rest Will Be Familiar to You from Cinema, Men Asleep*. London: Faber and Faber, 2019.

Foucault, Michel. "Of Other Spaces." *Diacritics* 16.1 (2014): 22–27.

Haß, Ulrike. "Chor." *Metzler Lexikon Theatertheorie*. Ed. Erika Fischer-Lichte, Doris Kolesch, and Matthias Warstat. Stuttgart: Metzler, 2014. 50–53.

Hauthal, Janine. *Metadrama und Theatralität: Gattungs- und Medienreflexion in zeitgenössischen englischen Theatertexten*. Trier: Wissenschaftlicher Verlag Trier, 2009.

Illouz, Eva. *Der Konsum der Romantik*. Frankfurt a. M.: Suhrkamp, 2007.

Moylan, Tom. "'Look into the Dark': On Dystopia and the Novum." *Learning from Other Worlds: Estrangement, Cognition, and the Politics of Science Fiction*. Ed. Patrick Parrinder. Durham: Duke UP, 2001. 51–71.

Nachtwey, Oliver. *Die Abstiegsgesellschaft: Über das Aufbegehren in der regressiven Moderne*. Frankfurt a. M.: Suhrkamp, 2018.

Nester, Nancy L. "The Empathetic Turn: The Relationship of Empathy to the Utopian Impulse." *The Individual and Utopia: A Multidisciplinary Study of Humanity and Perfection*. Ed. Clint Jones and Cameron Ellis. Farnham: Ashgate, 2015. 117–132.

Poschmann, Gerda. *Der nicht mehr dramatische Theatertext: Aktuelle Bühnenstücke und ihre dramaturgische Analyse*. Tübingen: Niemeyer, 1997.

Sargent, Lyman Tower. "The Three Faces of Utopianism Revisited." *Utopian Studies* 5.1 (1994): 1–37.

Sargent, Lyman Tower. *Utopianism: A Very Short Introduction*. Oxford: Oxford UP, 2010.

Sierz, Aleks. *The Theatre of Martin Crimp*. London: Bloomsbury, 2013.

Voigts, Eckart. "Introduction: The Dystopian Imagination – An Overview." *Dystopia, Science Fiction, Post-Apocalypse: Classics – New Tendencies – Model Interpretations*. Ed. Eckard Voigts and Alessandra Boller. Trier: Wissenschaftlicher Verlag Trier, 2015. 1–11.

Maria Marcsek-Fuchs
"Let the Doors Be Shut upon"... COVID-19: Relocating the Globe Theatre Stage to the Net

1 Intro: Let the Doors Be Shut

Let the doors be shut upon... COVID-19. This is exactly what happened at London's Globe Theatre[1] on 18 March 2020; but with this, the doors were equally shut upon all the actors, creative teams, and also on audience members, which meant a loss of 95 per cent of the entire income necessary for survival. The headlines of leading newspapers sounded accordingly dramatic, reading "Coronavirus could mean the final curtain for Shakespeare's Globe Theatre" in the *Independent* (Booth) or "Shakespeare's Globe Theatre may be another Coronavirus Casualty" in *Forbes* Magazine (Kelleher). Robert Myles, creator of the project *The Show Must Go Online*, which started "staging" all of Shakespeare's plays on Zoom only few weeks after the national lockdown was called out, stresses Shakespeare's central role in times as these:

> Being from working class origins, I believe Shakespeare is for *everyone*. Being a time of crisis and isolation, I believe it is more important than ever. Look no further than Nelson Mandela, and the Robben Island Bible. (Myles)

2 Utopian vs. Dystopian View

However, is Shakespeare at the Globe Theatre attainable for everyone? A utopian view on the institution would agree: tickets in the yard amount to only 5 pounds. The Globe's educational programme is designed to cater to all ages, classes, nationalities, and educational backgrounds, ranging from kindergarten to university levels, and its online sources are easily available, mostly for free. However, I would also argue that the key audiences are white, middle-class and, most importantly, have the funds to travel. Furthermore, the Globe as an "independent charity of education, performance and research" generates "95% of its income

---

[1] The official title of the theatre is Shakespeare's Globe. However, I will intermittently be using the short version: "the Globe," as is also customary in the theatre's own publications.

https://doi.org/10.1515/9783110758252-016

from theatre tickets and other on-site activities" (Shakespeare's Globe, "Celebrating"). This means, everything must be monetized through anything that can generate income. Thus, the Globe can be viewed as both a victim and an agent of neo-liberal politics and economics. While the entire Globe project serves as a communal performative space to engage with Shakespeare and his works, we may indeed fear that by losing "£2 million pounds a month" (Shakespeare's Globe, "Donate"), the survival of the theatre and the heritage site will depend on donations and governmental subsidy as well as on increasing the marketed revenue once the doors open again; despite the generous funding to come. This in turn can influence dramaturgical choices and educational programmes, even despite the trust's "primary purpose [being] to promote, maintain, improve and advance education" (The Shakespeare Globe Trust 10).

3 Shakespeare's Globe as "Experiencescape," "Nostalgiascape" and "Place of Imagination"

Notwithstanding the commercialization of popular Shakespeare(s) even before the pandemic, the democratic feel and the festive engagement with Shakespeare through performances, workshops, exhibitions, and guided tours invite audiences from around the world, bringing together those who are there for touristic reasons, those who enjoy the entertainment value, and those who strive to learn about Shakespeare. Many visitors are there for all these reasons combined. Right from the start, the reconstruction of "Shakespeare's Globe" was described as "a radical theatrical experiment," "cultivat[ing] intellectual curiosity and excit[ing] learning to make Shakespeare accessible for all" (Shakespeare's Globe, "About Us"). As Carson and Karim-Cooper point out, "[t]he Globe project began both as an educational and a theatrical endeavour" (Introduction 6). Aside from the Globe being a hub for artists, and researchers alike, the audience's fascination with this theatre experiment can be described by the following concepts borrowed from tourist studies: A visit to the theatre functions as an "experiencescape" (O'Dell), "nostalgiascape" (Gyimóthy) and "a place of imagination" (Reijnders). What inspires visitors is not the Shakespearean performance alone, but the entire experience connected to the visit: the architectural atmosphere, the music, the dancing, the queuing, 3 hrs of standing – all of it being a multi-sensual, multi-modal community event, involving what John Urry and Jonas Larsen call "bodily sensations and affect" (190). They point out further that "tourists have become bored of being mere spectators [and that they] are not only audiences but also performers" (Urry and Larsen 191). In this way, the visitors "are not just written upon, they enact and inscribe places with their own stories" (193), an observation which I think not only applies to travellers but also to audience members at the Globe complex. For this paper, I read Shakespeare's Globe in London as a performative space, allowing for an immersive engagement with anything Shakespearean: this involves focussing on both Shakespeare's works and on the simulacrum of a Renaissance stage, as well as the constructed narratives of past and present reception surrounding it. While the Globe's Shakespeare productions have been little immersive in allowing for audience participation, the experience at the Globe is. At the same time, the joy of simultaneously imagining and living an illusion of Shakespeare's past can be tackled with Szilvia Gyimothy's notion of "nostalgiascape," "a strategically stylized past." She describes further that "[n]ostalgic experiencescapes or nostalgiascapes may only work to full effect [...] if visitors develop a cognitive and emotional attachment to the place, era or people within the topical context" (113).

Yet, what past are we talking about here? Shakespeare's New Globe is a reconstruction based on Sam Wanamaker's initial dream, his tireless effort to struggle with city officials, plus many voices and stories, representations and research endeavours that add up to the simulacrum that we visit today as Shakespeare's Globe.[2] Stijn Reijnders uses the term "places of imagination" to describe the fictive result created by artists, which draw fans to search for the physical places that in turn inspired their fictional representations. For Reijnders,

> [p]laces of imagination are material reference points like objects or places, which for certain groups within society serve as material-symbolic references to a common imaginary world. Imagination for this purpose can be defined as a mental conception of an object, person or event that at a certain time and in a certain place is not actually present. Memory in this context is nothing more, or less, than the imagination of an event from the past. (14)

A result of such a physical place based on a fictional work is the real 221B Baker Street in London, housing Sherlock Holmes's living room in the museum. Although it is not a fictional place, I would still read the Globe Experiment as founded on such imagined places. At the same time it permanently (re-)creates new ones by all the connected projects, with the performances of Shakespeare's plays at their core, which in turn draw audiences to return to the Globe repeatedly to visit the physical embodiment of stories collected from educational, artistic and fan-generated content.

---

2 For a detailed study on *Shakespeare's Globe: A Theatrical Experiment* see Carson and Karim-Cooper as well as Vanessa Schormann's work on the New Globe Theatre as reconstruction. For further reading also consult Barry Day, and J. R. Mulryne and Margaret Shewring's edited collection.

4 Digital Stages

And then came COVID-19, the global pandemic, and governmental lockdowns. On 18 March 2020 the Globe theatre had to close all its doors, and with this at the same time decisive, responsible, and devastating decision everything went missing: the physical space, the artists, and the audience. The Globe theatre, and mostly our memory of our last visit had now to become the "imagined place," and the stage had to move to the world wide web, becoming a digital space of global and communal performance. Furthermore, several shifts had taken place: immersive experience turned into imagined "experience*space*." What had been central to the theatrical experiment until then, namely the live performances and the educational workshops in the Globe's on-site studios, was now side-lined to "a seat in [the] distracted globe" (*Hamlet*, 1.5.96–97) of our memory or longing.[3] And, what originally posed as some additional treat, paratext, and economic marketing strategy – like blogposts, podcasts, Instagram stories, *Facebook* announcements, and educational online resources – now turned into the main act; always accompanied and shaped by Shakespeare quotes.

In the following, I will show how the lockdown at Shakespeare's Globe evoked processes of relocation and the blurring of hierarchical, medial, and cultural boundaries through the move of all creative activity to the internet. The following discussion will highlight some of the ways in which this relocation of all activities, ranging from stage productions, exhibition tours, educational workshops, and academic festivals, has furthered global inclusion and foregrounded engagement with topical issues in politics, society and culture.[4]

---

[3] For a detailed description of the manyfold challenges the educational department had to face, see Karim-Cooper and Cuthbertson's essay on "The Creative Class."
[4] Shakespeare's Globe has offered a great abundance of ways to engage with Shakespeare for audiences and learners of all ages and has been orchestrating online resources to accompany the productions since launching their website and participating on social media via their own channels. However, the year 2020 has seen an intensified activity and a variety of formats in both furthering the existing features and creating new ways to connect with past and future audiences, all of which is orchestrated via several platforms. Since it is impossible to do justice to the complexity and detail of the Globe's move to the digital stage, this paper can only highlight a few results based on a selected number of examples.

4.1 Staging the Lockdown

The devastating news of closing doors on 18 March 2020 was simultaneously published on the Globe Theatre's website, blogpost, and all related social media platforms. Similarly to transmedia storytelling strategies,[5] each medial representation catered to the respective cognitive and emotional needs, but also helped to endure the lockdown with all its uncertainties, setbacks, and sorrows as a shared experience. For example, the Instagram post on closing day shows Michelle Terry (Artistic Director), Patrick Spottiswoode (Director of Globe Education), and Neil Constable (Chief Executive) standing in the empty yard of the theatre and holding a poster with a slightly changed line from *Julius Caesar* (5.1.128). It reads "[w]hen we do meet again, why, we shall smile," cleverly exchanging the word "if" in the play with the hopeful "when." While the blog post of the day immediately stressed the existential crisis of the theatre by asking for donations, the corresponding publication on Instagram started and ended by addressing both the global community and each individual viewer. Right from the start, it established a connection with "[h]ello from one Globe to another" and ended like a personal letter "[w]ith love, solidarity and thanks to you all from the Globe family" (the_globe, "Hello"), then repeating the Shakespeare quote. Only a few days later, first ways to stay connected were published: the blog, the Globe Player, and the theatre's own podcast, entitled *Such Stuff*, all of which had been available before but now focused on the crisis. Emotional responses on Instagram followed instantly, as did donations.[6] Like the virus, the posts and conversations went viral.

Yet unlike other theatres and Robert Myles's Zoom adaptations of the entire Shakespeare canon, the Globe refrained from creating full online production,[7]

---

[5] Henry Jenkins describes these as follows: "A transmedia story unfolds across multiple media platforms, with each new text making a distinctive and valuable contribution to the whole. In the ideal form of transmedia storytelling, each medium does what it does best" (97–98). While the digital presence and projects of Shakespeare's Globe Theatre are not designed to tell a fictional narrative that spreads across diverse media platforms, the audience's experience of the diverse digital formats, most of which mesh Shakespearean content with the real lockdown experience, on diverse digital platforms follows similar patterns and requires overcoming similar challenges on both the makers' and the recipients' sides.
[6] Donors ranged from organisations, such as Garfield Weston Foundation issuing the Western Culture Fond, to alumni of the MA Shakespeare Studies programme at King's College, London, whose "Read for the Globe" raised more than £13,000 (King's College London).
[7] For a comprehensive overview of online Shakespeare productions during lockdown see the special section in *Cahiers Élisabéthains* 103.1 (2020) and the introduction by its editors (Peter, Valls-Russel, and Yabut).

except for "*Macbeth: A Conjuring*: a socially distanced staged reading of *Macbeth* from the 2018 ensemble production" at the Sam Wanamaker Playhouse, which was reorganised into a film studio for the production of online content (The Shakespeare Globe Trust 12). Instead, they started a series of smaller formats and relocated the stage, performer, and topic to a web of orchestrated, communal Shakespeare experiences. On Instagram, these ranged from short photo reminiscences of past productions and tile panoramas showing the beauty of the theatre's architecture to short video readings.

The stage transferred to the world wide web, and with this, not only expanded the architectural confines to reach audiences across the world, but also blurred the boundaries between performers and viewers as well as between front- and backstage. The online features also relocated Shakespearean words into the topical contexts of the pandemic and turned the individual experience of lockdown into a communal one of empathy. While any Shakespearean production is chosen and directed in a way to reflect on urgent issues of the present, the relationship between the plays and 21st-century issues changed in features like the *Thought of the Week*, *Love in Isolation* and the Globe's contribution to the Mental Health project *Wander*.

In the following, I will trace these movements – the re-location of stages, the re-casting of performers, and the re-directing of topics – to investigate the rich creative potential of the digital alternatives found by Shakespeare's Globe during the lockdown in 2020. At the same time, I will trace the many ways in which these online offers helped invite an even wider audience to enjoy Shakespearean content.

4.2 Relocating Stages

In pre-COVID times, the Globe and the Sam Wanamaker Playhouse were the main stages of production and creativity. Blog, Podcast, Globe Player, and the theatre's social media platforms served as means to accompany the main feats of theatre and education. During the lockdown in 2020, however, these medial paratexts were now promoted to the main stage of attention and formed a mutually inspiring web of connecting points. At the same time, the physical stages turned into a void. If at all possible, the theatre stages as buildings were given new functions, such as broadcasting studios, lecture halls or spaces to welcome guests.[8] On the other hand, the buildings of both the Globe stage and the Sam

---

8 The Globe housed "two well-being *Spa Days* [...] in association with FMTW [for] nearly 100

Wanamaker Playhouse with their beautiful architecture and colourful interiors were promoted to "performers" in their own right through the many historical and contemporary photos and posts referring to the empty buildings.[9] Yet, the first posts to fill the empty spaces online, e. g. on the blog, were either educational (on Queen Elisabeth I) or helped establish a link to global events, such as Mother's or World Theatre Day. But, less than two weeks after the closing day, the Globe already announced four ways to "stay connected and share digital joy and wonder with [the] extended Globe family" (Blog post, 30 March 2020): 1) the new series *Love in Isolation*, 2) six *Free Globe Player Releases*, 3) *Educational Resources*, and 4) season five of the podcast *Such Stuff*, with some continued, refashioned or newly devised formats. Thus, Shakespeare-related content and further Globe activities spread over several media platforms and took on diverse medial shapes at the same time, all inviting audience participation more globally and differently than possible with the physical stage experience. The orchestration of these activities took on the quality of transmedia storytelling products in the following sense: On the one hand, their interplay needed to be designed in a way that the activities (main text) and their social media presence (paratext) were to become one but also cater to different audiences through different online activities (ranging from plays to workshops, and from university courses to storytelling for toddlers). On the other hand, the audience needed to acquire skills in transmedia story telling reception by being able to handle the great abundance of offerings and the wide variety of media platforms (ranging from blog posts, podcasts and YouTube channels to the different social media apps) without losing track.

One of these globally and multimodally engaging Shakespeare experiences was the watch parties on the nights when the Globe streamed their past productions on YouTube.[10] One hour prior to the performance, audience members were

---

freelance artists" and welcomed other formats to their stages, such as concerts (e. g. the BBC singers) and Strictly Come Dancing (The Shakespeare Globe Trust 12).

9 Some *Facebook* posts of Shakespeare's Globe meshed the history of the original Globe with that of its twentieth-century reconstruction (*Facebook*, 29 June 2020). Others addressed the building with a panoramic photo of the empty space accompanied by the quote "O beauty, / Till now I never knew thee" from Shakespeare's *Henry VIII* (*Facebook*, 29 July 2020). Yet others revealed the Globe architecture as being included in Helen Murray's photo series *Our Empty Theatres*, with the photographer's question: "What are these buildings, without all the glorious, wonderful, different kinds of people that make these spaces?" (*Facebook*, 23 July 2020).
10 The six plays included into this generous offer of streaming plays for free were *Hamlet* (2018), *Romeo and Juliet* (2009), *The Two Noble Kinsmen* (2018), *The Winter's Tale* (2018), *The Merry Wives of Windsor* (2019), and *A Midsummer Night's Dream* (2013), the first and last starring Michelle Terry.

invited to join the introductions and discussions on Twitter as well as the live chat during the play. Through this, centre stage and audience space were meshed. Actors speaking on screen and audience members commenting on the action happened simultaneously. On *Facebook,* audience members of these online streams were even addressed like actors backstage: Messages like "It is your one hour call," announced the "semi-staged reading of the *The Scottish Play* [...] in the Sam Wanamaker Playhouse," explicitly requesting comments alongside the performance on Bonfire Night (Shakespeare's Globe, "Something").[11]

Many blog and social media posts allowed the audience backstage by offering photo and video material of areas otherwise unreachable. While a peek into the rehearsal room had existed before the pandemic as a special treat, this feature was assigned a more central role in the Globe experience during the lockdown. It was now incorporated into the web of communal activities joining the creative team and the viewers on the same digital stage. Furthermore, while "behind the scenes"-material used to wet the appetite for the on-site production to come, posts on the blog accompanying the "YouTube Premiere of *Hamlet*" (Shakespeare's Globe, "Get") now invited a nostalgic view back to the 2018 production, allowing a more detailed view and more general insights into the working processes of Globe productions in general. Yet, the post meshes Shakespeare and COVID-related content in one blog and switches swiftly to the present lockdown. Already the subtitle of the second section highlights the devastating situation of the entire theatre world by claiming that "[w]e face the most challenging time in history as an industry," immediately offering a link to the donation section of the Globe's website (Shakespeare's Globe, "Get"). As with most of the posts, Shakespeare quotes function as a catalyst for the arguments.

> "In 1599, when Hamlet stood and uttered the words 'Now I am alone' – he would have been surrounded by up to 3,000 people," says Artistic Director, Michelle Terry. "Now we are alone, but we are also in the company of billions, from all around the globe, finding the most inspiring ways to be alone, together." (Shakespeare's Globe, "Get")

Many of the statements are at the same time self-revealing and reaching out to the community. Through this, the posts create an atmosphere of "we are in this together." Thus, the next subheader reads "[w]e also want to ensure that our heart

---

[11] This screening was part of the Shakespeare and Fear Festival, created directly for Halloween. It was one of the many examples which orchestrated entertainment through the dramatized reading with educational formats: *In Conversation: Fear in Our Moment* was a "panel discussion looking at the power and potency of fear in our time and throughout history" (The Shakespeare Globe Trust 12).

remains open," re-confirming their mission "of making Shakespeare accessible to all." Posts like these also included overturning theatrical and dramaturgical hierarchies: while the main stage was closed, backstage became frontstage, and props and costumes were presented as visual stars on Instagram: From 22 to 23 September 2020, the Globe offered "a sneaky glimpse into [their] costume and props stores [...] including boxes of party animals ... and tennis balls (thanks France ☺)" (the_globe, "Exit"). Shakespeare references and jokes kept the addressees actively engaged. The post on 25 September directly invited audience responses: It shows a hobby horse, easily mistaken for a donkey's head, standing among brooms in a storage space. This image is accompanied by the lines "What do you see? You see an asshead of your own, do you," followed by "[b]less thee Bottom, have you been in our Props store this whole time?" A fan responded with "A horse, a horse for my kingdom. Wait a minute. I just gave up my kingdom for a hobby horse" (the_globe, "What do you see?"; comment by grehudson8088). Audience members thus became co-authors of the witcombat on the digital stage.

The Globe's "experience*scape*" transformed into a digital and weblike "experience*space*" for all. Notions of "nostalgiascape" and "place of imagination," as associated with the Globe, have now changed locations and features: nostalgic feelings have moved from playful engagement with some distant, ungraspable past of the Renaissance and its iconic author to a very real, personal sensation of longing and loss during the lockdown – a longing that looks back at lived experiences at the Globe or at those trips that could not (yet) be made. Thus, online features like the countless images and short clips of past productions, the stages' architecture and audiences in the yard now stand in for the lived experience. The *Facebook* post on 11 August 2020 reads: "You're escaping to a desert island ... which of Shakespeare's works out of his plays, narrative poems and sonnets are you taking with you?," and earlier, on 25 June 2020: "If you could be watching any play right now in our wooden 'O' on this glorious sunny day (pint in hand [...]), which would it be?" While posts like these both represent and incite feelings of nostalgia and longing, they also help imagine not only the play's performance but also the bodily experience of being embraced by both audience and architecture at future theatre visits.

4.3 Re-Casting Performers

Moving the real stage to the digital space also brought about a change in casting: Fans were incorporated into professional projects. Accordingly, directors and filmmakers collaborated side by side with amateurs and fans in producing video content for the Globe. By the same token, the architecture of the Globe Theatre itself – a non-human agent in the sense of Latour's actor-network theory – became a star of the web output.

The following examples show how Shakespeare's Globe participated in national and transnational projects, and thus entered the global web of digital creativity during the lockdown. Often, these projects were inspired by well-known brands, newspaper agencies or global movements. The first of the three examples in this paper is *Wander*,[12] a project initiated by Ribena in consultation with the Mental Health Foundation. The video filmed at the Globe is part of "an uplifting series of films [using cognitive film-making techniques] designed to bring a bit of creative inspiration, escapism and meditation to everyone during these difficult times." The aim was to "help us alleviate the stress we're all feeling during lockdown, provide professional guidance to help deal with anxiety, and at the same time inspire and encourage everyone to express their creativity" ("Welcome to Wander").[13] The contribution at Shakespeare's Globe presents a mash-up of present-day images and past productions at both the Globe and the Sam Wanamaker playhouse. It shows action on- and backstage, starring actors, crew, and audience members at the same time. Furthermore, several productions from both the distant and the recent past are juggled into a mix of powerful action and silent speakers, because the only voice we hear is Joseph Marcell[14] performing Brutus's monologue from Shakespeare's *Julius Caesar* (2.1.113–139), perfectly set to the melancholic score.

The video opens with a nightly glance at the Thames followed by several perspectives on the Globe theatre and a breath-taking crane shot onto the stage. It is very telling that the makers chose impressions of deserted streets and closed doors opening onto empty stages, expressing both calmness and loneliness. Images of the abandoned yard are contrasted with those of packed audiences from pre-COVID productions, invoking feelings of nostalgia and long-

---

[12] This particular video filmed at the Globe and directed by Danny Cappozi is part of a larger series produced by Beau Kerouac and Willow Moon Productions in collaboration with the Mental Health Foundation. It was aired on 19 May 2020: https://www.shakespearesglobe.com/discover/blogs-and-features/2020/05/19/welcome-to-wander/

[13] The Globe Theatre's contribution to the project is one in a series, showing iconic cultural spaces and canonical readings performed by renowned stars, such as Richard E. Grant reading from Carroll's *Alice in Wonderland* at the V&A Museum or Juliet Stevenson performing poems by Keats at the English National Ballet. Each video demonstrates the merits of the institution and the artist as well as highlighting the loss we are experiencing during lockdown ("Welcome to Wander").

[14] Joseph Marcell is an acclaimed actor, having starred with the Royal Shakespeare Company and *Fresh Prince of Bel-Air*. He is now a board member at Shakespeare's Globe.

ing. This impression is enhanced by the choice of productions incorporating all genres as well as mixing Shakespearean texts with new writing, such as Jessica Swale's *Nell Gwynn*. Thus, the video addresses everyone in the Globe family, ranging from stars and backstage professionals to volunteers and audience members. What adds to the strong message of the video is the monologue spoken by Joseph Marcell. Together with the images it can be read as an insistent call for resilience in times of lockdown, furlough, and isolation: Brutus claims in this speech that no oath is needed to rise up to action, other than empathy with the suffering, a communal bond, and the belief in the common cause – statements that can easily be transferred to the present situation at the Globe. While one is watching passionate faces and actions on stage, the Shakespearean lines invoke a sense of pity: "If not the face of men, / The sufferance of our souls, time's abuse, / If these be motives weak, break off betimes, / And every man hence to his idle bed" (2.1.113–116). The "high-sighted tyranny [that] rage[s] on / Till each man drop by lottery" (2.1.117–118) mentioned by Brutus can be related to the virus and the economic crisis straining both the theatre and all artists.[15] The parallels between Shakespeare's lines and the present situation culminate when both text and music reach their climax: Joseph Marcell recites "do not stain / The even virtue of our enterprise, / Nor th'insuppressive mettle of our spirits, / To think that or our cause or our performance / Did need an oath" (2.1.131–135). This video was published both on the Ribena project page and on the Globe's blog, the latter adding Shakespeare's lines as subtitles to the video, thus foregrounding the power of Shakespeare's words.

The second example highlighting the authority of Shakespearean texts and the power of audience participation during lockdown is the Globe's video project *Love in Isolation*, a feature newly designed to replace the missing productions on the physical stage. For this project, Shakespeare's Globe

> partnered with *The Guardian* [...] to relaunch their Shakespeare Solos series. Originally created in 2016 to mark the 400th anniversary of Shakespeare's death, this is a video series of leading actors performing some of Shakespeare's greatest speeches. For this new collaboration, *The Guardian* invited the general public to join leading actors in performing three of Shakespeare's iconic speeches from *As You Like It*, *Hamlet* and *The Tempest*, [sic] their place of self-isolation. Over 500 people from around the world submitted and a selection of those performances have been edited together here as part of the relaunch of Shakespeare Solos. (Shakespeare's Globe, *Love in Isolation*)

---

15 For a study on "viral theatre" interrelating literary and medical studies as well as giving comprehensive insights into the changes that take place as performances are created for Zoom, see Heidi Liedke and Monika Pietrzak-Franger.

Not only did the videos include selections from all genres of the Shakespeare oeuvre and celebrities both on and off the stage, such as Steven Fry, Marcello Magni, Mara Allen, or Athena Stevens, but also members of the public of all ages. The video accompanying the relaunch of the "Solo Series," which premiered on Shakespeare's birthday, fittingly bore the title "Three Speeches Performed by the Quarantine Players." Yet while the solos were staged "in [the speakers' own] places of solitude and isolation" (Shakespeare's Globe, "Celebrate Shakespeare's Birthday"), the opportunity to perform alongside stars offered communal experiences, inspired fan-based creativity and with that helped each participant contribute their own "place of imagination" by setting the respective line in a garden or on the couch in their private homes. By letting the speakers choose their own environment, prop and angle, the lines from these speeches meshed performance with personal message.

The third example of connecting Shakespearean content with the lockdown experience, thus emphasizing both the project's topicality and the theatre's engagement in current affairs, is the Globe's contribution to the Refugee Week in 2020. Shakespeare's Globe invited "Compass Collective," an endeavor helping young refugees participate in theatre, music, and film projects, to add creative input to the *Love in Isolation* series. The selected speeches were taken from iconic plays such as *King Lear*, *Romeo and Juliet*, *Hamlet*, and *As You Like it*, which easily and very touchingly helped make the connection to migrant experiences of isolation, banishment, confinement, and exile.

4.4 Re-Directing Topics

As these examples have demonstrated, relocating the stage to the world wide web and recasting performers to come from all walks of life also resulted in foregrounding topical issues of our times in general and during lockdown in particular. Shakespeare's Globe answered to present challenges like remote teaching through their engagement in online education, to coping with anxieties through the Shakespeare and Fear Festival, to continued gender discrimination by equalising cast and production choices, and to racism though their participation in the "Black Lives Matter" movement by introducing an Anti-Racist Taskforce, "author[ing] an anti-racist statement of intent and action. [...] Shakespeare's Globe made a public commitment to become an anti-racist, pro-equality organisation" (The Shakespeare Globe Trust 12). Not only were all of the existing formats of performance, education and discussion "pivot[ed] to digital platforms" (The Shakespeare Globe Trust 5), but by having to move online, educational and academic discourses on topical issues were brought centre stage. This highlighted the Globe's endeavour to both reread Shakespearean content through the contemporary lens and participate digitally in the global discourses around the most challenging topics of our times.

In the following, I will shed light on the way topical issues were taken centre stage through digitalisation by zooming in on three selected examples: a) The Playing Shakespeare with Deutsche Bank production of *Romeo and Juliet* – gone transmedial, b) Shakespeare and Race Festivals – gone digital, and c) Shakespeare's Globe – gone personal.

a) Playing Shakespeare with Deutsche Bank: Gone Transmedial

To help remote teaching during the pandemic in 2020, the Globe's educational department offered free YouTube screenings and watch parties of their 2019 Playing Shakespeare production of *Romeo and Juliet*. These shortened versions, accompanied by a rich array of educational resources, are aimed at secondary school students. This was a follow-up after the online success of streaming a similar youth version of *Macbeth* earlier that year, which was watched by 470,000 viewers from May to July. Neither the Playing with Deutsche Bank series nor holding educational workshops in addition to the event were new. Moving everything online and conducting workshops with children via Zoom, however, was. Shakespeare's Globe intensified the orchestration of the accompanying material, ranging from online resources and workshops for teachers (CPD programme) to family formats for ages 9–12, all in connection with this streaming event. Since everything was only available online, the Globe's educational formats, resources and topical issues gained more public attention. With the Shakespeare and Race Festival having recently finished when the youth version was streamed, students still had the opportunity to follow the discussion on race and mental health via the feature called "Behind Closed Doors: *Romeo and Juliet*."

b) Shakespeare and Race Festival 2020: Gone Digital

The company which was going to perform the full play about the "star-crossed lovers" live during the Globe's summer season in 2020 met in "an honest conversation about race, beauty, femininity and mental health, [...] consider[ing] their roles as artists and assess[ing] the impact of Shakespeare's language on audiences today" (Shakespeare's Globe, "Behind Closed Doors"). This format was part of the Shakespeare and Race Festival, now in its third series, but gone digital for the first time. Not only did this involve changing the Sam Wanamaker

Playhouse into a broadcasting studio with the help of WarnerMedia (The Shakespeare Globe Trust 5), but it radically changed both the function of the indoor theatre and the performance strategies of all on stage. The Shakespeare and Race Festival also included *Notes to the Forgotten She-Wolves*, lending a voice to women of colour, and the *Conversation: Reckoning with our Past*. By digitising and even more so conceptualising these topical festivals as online films and webinars, they not only reached a global audience and were accessible for a longer period. These formats also demonstrated globally how an engagement with Shakespeare can be used to discuss more universal issues, especially the ones pressing at this time of lockdown. In "Behind Closed Doors: *Romeo and Juliet*" director Ola Ince and actors Alfred Enoch and Sargon Yelda were joined by psychotherapist Rachel Williams to discuss the play in the light of mental stress and anxiety during the pandemic. As Williams claims, "Romeo does not ever get to sit with his pain," when he finds out that he is banished – a problem that lies at the heart of modern health problems in her view. In this vein, the panellists meshed their readings of the play with medical information, and their own lockdown experiences, by that connecting Shakespeare's lines to young people's present condition. Similar to this, the *Conversation: Reckoning with our Past*, opens up Shakespearean content to provoking questions in times of the "Me Too" and "Black Lives Matter" Movements and in response to the killing of George Floyd: "Professor Farah Karim-Cooper [founder of the Shakespeare and Race Festival in 2018, was joined] by novelist and academic Preti Taneja, Historian and President of the Royal Historical Society, Margot Finn, and actor and director Elliot Barnes-Worrell to discuss British history, the colonial past, racial identity and how best to tell [and challenge] our collective stories" (Shakespeare's Globe, "In Conversation"). This past also included the provocative and postcolonial rereading of Shakespeare's works themselves, at the same time opening up to global challenges of the present.

c) Shakespeare's Globe: Gone Personal

In the first moments after the lockdown had been announced, the formats resulting from it brought the universal pandemic to the Globe and staged this experience online by zooming in on the individual experiences of isolation and insecurity, thus invoking a sense of identification and communal empathy. *Thought of the Week* was one of these formats introduced immediately after the lockdown had started. Michelle Terry published a short text weekly for the entire time until the doors reopened in spring 2021, mirroring both the Globe's stance on isolation and the very private self of the reader. The Globe's website describes the format as follows:

> Each week during the UK's current Coronavirus crisis, our Artistic Director Michelle Terry shares her thought of the week. Using Shakespeare's language, Michelle reflects on the individual and universal meaning of the words. By giving personal and emotional insight, she uses the quote to relate to, and express, the mood of this uncertain time.

All entries worked with a single Shakespeare quote to tackle one of the many problems during the pandemic, making the texts feel like a personal message to each individual reader. For example, the entry "Hope and Help" centred around a quote from *The Merchant of Venice:* "How far that little candle throws his beams" (5.1.99). Michelle Terry uses this line to highlight the way a good deed can shine in a weary world. She links the quote to an incident she experienced when courageous bystanders at a grocery store helped an elderly lady with her heavy bags and protected her against the selfish complaints of queuing customers. The subtitle of this post reads "Artistic Director Michelle Terry shares a feeling of hopefulness as she witnesses the kindness of the human spirit" (Terry, "Hope"). The titles of the series reflect the journey from lockdown to reopening, ranging from "loss and lament" and "power of the mind" to "readiness is all" and "new journeys." The format ended shortly before the reopening of the Globe on 16 March 2021 with the telling title and quote "What's Past is Prologue" from *The Tempest*. In this last message, Terry characterises these times as "the most disorientating, devastating, uncertain, frightening [...] in recent history, where the individual and the collective faced existential crisis after existential crisis." By enumerating the challenges, she makes a strong statement for diversity, inclusion, and anti-racist strategies in pointing out that "[u]nconscious bias training," "[a]nti-racist training," and "diversity recruitment will not be enough if the systems within which we all operate remain the same" (Terry, "What's Past"). She calls for consistent change at Shakespeare's Globe at large and within each production process. This also includes her thoughts on "how this new digital space can explode, explore and support the work and reach people" (Terry, "What's Past").

5 Contradictions, Silver-Linings and Visions

The rich outcome of this year-long lockdown at Shakespeare's Globe offers two contradicting perspectives. The utopian/optimistic view sheds light on the many ways in which the theatre's move to the digital space offered new collaborative stages for more global and participatory ways of engaging with Shakespeare. Not only were the productions available to everyone, but the global public was also invited to the "wooden O." Moreover, personal and global issues were tackled simultaneously. As the Strategic Report and Review of 2019/2020 reveals, "the Globe Player free series secured an incredible 2.7 million views worldwide, from the UK to the USA, India to Australia" (The Shakespeare Globe Trust 4). Likewise, the virtual tours of the Globe reached "students and teachers across the world," including members from Shanghai to Greece, which also meant opening up of digital "experience*spaces*" for the more topical formats. Furthermore, in the first days when the physical stage allowed guests, free tickets were offered to NHS workers, and "'pay what you can' timeslots" were introduced to increase accessibility for all (The Shakespeare Globe Trust 6, 13).

Yet, the more dystopian perspective, on the other hand, reveals the devastating damage that has resulted from the lockdown, such as having to let go many esteemed colleagues, freelancers, and volunteers, and losing 95 per cent of the income, possibly facing permanent closure. Had it not been for the "lifeline grant of just under £3 million from the Government's Culture Recovery Fund" and later for the loan applied for in the second round of the scheme, plus the many donations worldwide, Shakespeare's Globe could not have opened for a summer season in person in 2021 (The Shakespeare Globe Trust 7).

This leads to two contrasting questions that remain unanswered: How can Shakespeare's Globe help to enjoy live performances, and by filling the theatres, archives and on-site workshops not only bring back Shakespearean experience-*scapes*, but also recover financially in order to survive? At the same time, how can the Globe hold on to the newly formed ties to the global community and newly found ways of sharing Shakespearean content digitally for all? In short, how to combine economic constraint with the digital democratisation of Shakespeare? The Globe's plan is to combine "live and live-streamed productions, workshops and events [in order to] enable us to maintain meaningful contact with our audiences, even those who will have been most isolated" (The Shakespeare Globe Trust 14). Let us hope for the best, so that Shakespeare at the Globe can remain a cultural benefit to all, also including audiences from socially, culturally and regionally disadvantaged backgrounds. Alongside the many crises and constraints which the lockdown brought to Shakespeare's Globe, closure also opened up a new path towards post-COVID theatre. This time illustrated the great potential of digitalising our "imaginary forces" (*Henry V*, Prologue) and of realising that "all the world [wide web]'s a stage" (*As You Like It*, 2.7.139), both on-site and on-line.

Works Cited

Booth, William. "Coronavirus Could Mean the Final Curtain for Shakespeare's Globe Theatre." *Independent* 20 May 2020. Web. 31 Jan. 2022. <https://www.independent.co.uk/arts-entertainment/globe-theatre-shut-london-shakespeare-south-bank-coronavirus-a9523611.html>.
Carson, Christie, and Farah Karim-Cooper, Introduction. *Shakespeare's Globe: A Theatrical Experiment*. Ed. Christie Carson and Farah Karim-Cooper. Cambridge: Cambridge UP, 2008. 1–12.
Day, Barry. *This Wooden 'O': Shakespeare's Globe Reborn*. London: Oberon Books, 1996.
the_globe. "Hello from one Globe to another." *Instagram*. 18 March 2020. Web. 13 March 2022. https://www.instagram.com/p/B94d55GHQ0c/.
the_globe. "Exit, Pursued by a Bear". *Instagram*. 23 Sept. 2020. Web. 13 March 2022. https://www.instagram.com/p/CFesd0Bnd3e/.
the_globe. "What do You See?" *Instagram*. 25 Sept. 2020. Web. 13 March 2022. https://www.instagram.com/p/CFjwB7onzoF/.
Gyimóthy, Szilvia. "Nostalgiascapes: The Renaissance of Danish Countryside Inns." *Experiencescapes: Tourism, Culture and Economy*. Ed. Tom O'Dell. Frederiksberg: Copenhagen Business School P, 2010. 112–124.
Jenkins, Henry. *Convergence Culture: Where Old and New Media Collide*. New York: New York UP, 2008.
Karim-Cooper, Farah, and Lucy Cuthbertson. "Creative Class." *Globe* (Winter 2021): 48–53.
Kelleher, Suzanne Rowan. "Coronavirus Could Mean the Final Curtain for Shakespeare's Globe Theatre." *Forbes*, 19 May 2020. Web. 31 Jan. 2022. <https://www.forbes.com/sites/suzannerowankelleher/2020/05/19/shakespeares-globe-theatre-may-be-another-coronavirus-casualty/>.
King's College London. "Former King's Students Raise over £13,000 for Shakespeare's Globe," 29 May 2020. Web. 31 Jan. 2022. <https://www.kcl.ac.uk/news/former-kings-students-raise-over-13000-for-shakespeares-globe>.
Liedke, Heidi, and Monika Pietrzak-Franger. "Viral Theatre: Preliminary Thoughts on the Impact of the Covid-19 Pandemic on Online Theatre." *Journal of Contemporary Drama in English* 9.1 (2021): 128–144.
Mulryne, J.R., and Margaret Shewring. Eds. *Shakespeare's Globe Rebuilt*. Cambridge: Cambridge UP, 1997.
Myles, Robert. *The Show Must Go Online: Our Story*. Web. 31 Jan. 2022. <https://robmyles.co.uk/theshowmustgoonline/>.
O'Dell, Tom, ed. *Experiencescapes: Tourism, Culture and Economy*. Frederiksberg: Copenhagen Business School P, 2010.
Reijnders, Stijn. *Places of the Imagination: Media, Tourism, Culture*. London: Routledge, 2016.
Schormann, Vanessa. *Shakespeares Globe: Repliken, Rekonstruktionen und Bespielbarkeit*. Heidelberg: Winter, 2002.
The Shakespeare Globe Trust. *Annual Report and Financial Statements 2020*, 2020. Web. 7 Feb. 2022. <https://cdn.shakespearesglobe.com/uploads/2021/10/Shakespeares-Globe-Annual-Report-and-Financial-Statements-2020.pdf>.

Shakespeare's Globe. "About Us." Web. 31 Jan. 2022. <https://www.shakespearesglobe.com/discover/about-us>.

Shakespeare's Globe. "Behind Closed Doors: *Romeo and Juliet*." Web. 8 Feb. 2022. <https://www.shakespearesglobe.com/whats-on/behind-closed-doors-romeo-and-juliet-2020/>.

Shakespeare's Globe. "Celebrate Shakespeare's Birthday with Us." Web. 8 Feb. 2022. <https://www.shakespearesglobe.com/discover/blogs-and-features/2020/04/23/celebrate-shakespeares-birthday-with-us/>.

Shakespeare's Globe. "Celebrating the Support of Our Audiences Around the Globe." 2021. Web. 30 Nov. 2021. <https://www.shakespearesglobe.com/discover/blogs-and-features/2021/11/30/celebrating-the-support-of-our-audiences-around-the-globe/>.

Shakespeare's Globe. "Donate." Web. 5 Feb. 2022. <https://www.shakespearesglobe.com/join-and-support/donate/>.

Shakespeare's Globe. "Get Thee Ready for Hamlet on YouTube." 6 April 2020. Web. 21 April 2022. <https://www.shakespearesglobe.com/discover/blogs-and-features/2020/04/06/get-thee-ready-for-hamlet-on-youtube/>.

Shakespeare's Globe. *In Conversation: Reckoning with Our Past*. Web. 5 Feb. 2022. <https://www.shakespearesglobe.com/whats-on/in-conversation-reckoning-with-our-past-2020/>.

Shakespeare's Globe. *Love in Isolation*. Web. 8 Feb. 2020. <https://www.shakespearesglobe.com/watch/love-in-isolation-2020/>.

Shakespeare's Globe. "Something Wicked This Way Comes…" *Facebook*, 5 November 2020. Web. 26 April 2022. <https://www.facebook.com/ShakespearesGlobe/photos/a.205335300773/10159151467400774/>.

Smith J. Peter, Janice Valls-Russel, and Daniel Yabut. "Shakespeare Under Global Lockdown: Introduction." *Cahiers Élisabéthains: A Journal of English Renaissance Studies* 103.1 (2020): 101–111.

Terry, Michelle. "Thought of the Week: Hope and Help." 10 Apr. 2020. Web. 12 Feb. 2022. <https://www.shakespearesglobe.com/discover/blogs-and-features/2020/04/10/thought-of-the-week-hope-and-help/>.

Terry, Michelle. "What's Past Is Prologue." 16 Mar. 2021. Web. 13 March 2022. <https://www.shakespearesglobe.com/discover/blogs-and-features/2021/03/16/whats-past-is-prologue/>.

Urry, John, and Jonas Larsen. *The Tourist Gaze 3.0*. [3rd ed.]. London: Sage, 2011.

"Welcome to Wander: Walks Through Beautiful Spaces Accompanied by the World's Favourite Voices." Web. 10 Feb. 2022. <https://www.ribena.co.uk/creative-hub/inspiration/>.

Notes on Contributors

Vicky Angelaki is Professor of English Literature at Mid Sweden University, Sweden.

Elaine Aston is Professor Emerita of Theatre at Lancaster University, UK.

Sebastian Berg is Associate Professor of Social and Cultural Studies at Ruhr University Bochum, Germany.

Paola Botham is Lecturer in Drama at Birmingham City University, UK.

Matthias Göhrmann is a PhD student at the University of Passau, Germany, where he is writing his transdisciplinary doctoral thesis on the "culture war" and its representations in contemporary British and American theatre plays.

Dennis Henneböhl is based at Paderborn University, Germany, where he is currently in the final stages of his PhD project on nostalgia in contemporary British culture and society.

Maria Marcsek-Fuchs is Lecturer in English Literature and Cultural Studies at TU Braunschweig, Germany.

Anette Pankratz is Professor of British Cultural Studies at Ruhr University Bochum, Germany.

Nicole Pohl is Professor in Early Modern Literature and Critical Theory at Oxford Brookes University, UK.

Trish Reid is Professor of Theatre and Performance and Head of the School of Arts and Communication Design at the University of Reading, UK.

Julia Schneider is a Research Assistant at Paderborn University, Germany, and currently working on her PhD thesis entitled "Female Bodies in Transition: Liminality in Dystopian Young Adult Fiction."

Peter Paul Schnierer is Professor of English Literature at Heidelberg University, Germany.

Luciana Tamas is a Lecturer and Research Assistant at Braunschweig University of Art, Germany. She is currently working on two PhDs and is also a visual artist, curator and translator.

Merle Tönnies is Professor of English Literature and British Cultural Studies at Paderborn University, Germany.

Leila Michelle Vaziri is a Lecturer and Research Assistant in English Literature at the University of Augsburg, Germany. She is currently working on her PhD thesis entitled "The Theatre of Anxiety: Border Crossings in 21st-Century British Theatre."

Eckart Voigts is Professor of English Literature and Cultural Studies at TU Braunschweig, Germany.

Ilka Zänger is a doctoral candidate at Paderborn University, Germany, where she is writing her PhD thesis on dystopian elements in postdramatic theatre texts. She is working as theatre educator at Junges Schauspiel Düsseldorf.

Index of Names

Abramović, Marina   29
Adams, Emma   88
– *Animals*   167
– *Ugly*   88
Adorno, Theodor   23, 36
Agamben, Giorgio   4, 30, 33
Ahern, Cecilia   73–75, 78f., 83f.
– *Flawed*   73–84, 118, 120–122, 124–129
– *Perfect*   51f., 74f., 82f.
Ahmed, Sara   2, 7, 12, 16, 20, 23, 186–188, 190, 192, 194, 197
Alighieri, Dante   22, 105, 201, 203f.
– *Divina Commedia*   201f., 215
– *Inferno*   22, 201
Althusser, Louis   94
Anderson, Benedict   138

Baccolini, Raffaella   2, 4, 57–59, 61, 67–69, 152, 160, 209f., 221
Barker, Howard   59
– *That Good Between Us*   59
Bartlett, Mike   44, 88, 152f., 167f.
– *Charles III*   152f., 162
– *Earthquakes in London*   44, 88, 167f.
Bassett, Linda   20f., 23
Baudelaire, Charles   105
Baudrillard, Jean   216
Bauman, Zygmunt   158f., 161, 194, 198
– *Liquid Modernity*   158
– *Retrotopia*   159
Bean, Richard   7, 88, 165, 167–170, 172, 174, 176–180
– *The Heretic*   7, 88, 165, 167–173, 177–180
Beuys, Joseph   29, 37
– *7000 Eichen   Stadtverwaldung statt Stadtverwaltung*   29
Billington, Michael   20, 62, 65, 90, 94, 179, 191, 196
Birch, Alice   152, 154
– *Revolt. She Said. Revolt Again*   152
Blair, Tony   117, 127, 172
Blake, William   138, 203
Bloch, Ernst   4, 18f., 23f., 30

Blythe, Alecky   151
– *Little Revolution*   151
Bond, Edward   59, 78f., 158
– *Lear*   59
– *The Children*   78f.
– *The War Plays*   158
Booker, Marvin Keith   63, 160, 187
Bourdieu, Pierre   104, 106f.
Brecht, Bertolt   12, 34, 61, 94, 150f., 160
Brenton, Howard   3, 59f., 143, 151, 158
– *Bloody Poetry*   60
– *55 Days*   151, 153
– *Greenland*   3, 60, 158
– *Sore Throats*   60
– *The Churchill Play*   59
– *Thirteenth Night*   60
Brown, Gordon   117f., 174
Brown, Wendy   6, 88–91, 93f., 117f.
Buffini, Moira   61, 88
– *Greenland*   88, 167f.
Butterworth, Jez   151, 153
– *Jerusalem*   151, 153f., 159, 162
Butucea, Vlad   87, 97f.
– *Glowstick*   87, 97–99
– *Interference*   6, 87f., 90, 95–99

Cavendish, Margaret   3
– *The Convent of Pleasure*   3
Charman, Matt   88
Churchill, Caryl   2f., 5, 12, 16, 20–24, 44, 47, 50, 52, 57–63, 65–69, 83, 88, 98, 150, 152, 154, 167
– *Escaped Alone*   2, 12, 20f., 23, 44, 47, 50, 52, 60, 62, 88, 167
– *Far Away*   5, 20, 47, 50, 58, 60, 83, 152, 167
– *Here We Go*   98
– *Light Shining in Buckinghamshire*   3, 60, 69, 150f.
– *Owners*   5, 57, 60–63, 65–68
– *The Ants*   61
– *You've No Need to Be Frightened*   61
Claeys, Gregory   58, 73, 81

Index of Names

Coates, David   6, 117–129
Conrad, Joseph   202
– *Heart of Darkness*   202
Cooper, Melinda   89
Corbyn, Jeremy   127
Cowper, William   46
Crimp, Martin   5, 8, 43–54, 209–222
– *Attempts on Her Life*   209, 215, 220
– *Cruel and Tender*   210
– *Dealing with Clair*   210
– *Definitely the Bahamas*   210
– *Getting Attention*   210
– *In the Republic of Happiness*   209, 211, 214–218, 220–221
– *In the Valley*   5, 43–45, 47, 50f., 54
– *Play with Repeats*   210
– *The Rest Will Be Familiar to You from Cinema*   219f.
Critchley, Simon   188

Dolan, Jill   2, 4f., 22f., 35, 150
Duffy, Carol Ann   143, 162
– *My Country*   143, 162
Duffy, Clare   88, 167
– *Arctic Oil*   88, 167
Dyer, Richard   2, 13f., 16, 19, 21f.

Eccleshare, Thomas   167
– *Pastoral*   167
Eddo-Lodge, Reni   19
Eliot, T. S.   114
Etchells, Tim   109–113
Euripides   219
– *The Phoenician Women*   219

Farage, Nigel   140
Featherstone, Vicky   51
Fisher, Mark   4, 156, 158, 161, 177
Forced Entertainment   6, 109–111
– *Emanuelle Enchanted (or a Description of This World as if It Were a Beautiful Place)*   110, 113
– *End Meeting for All*   110
– *(Let the Water Run its Course) to the Sea that Made the Promise*   110
Foucault, Michel   74, 76f., 79–82, 84, 215
Fox, Liam   138

Giddens, Anthony   127
Gove, Michael   136f., 139
Grayling, Chris   136f.

Hall, Stuart   21f., 169, 217
Hancock, Matt   141
Haraway, Donna   30, 33, 39
Harris, Zinnie   7f., 185, 193, 195, 198, 205f.
– *How to Hold Your Breath*   7f., 185, 188, 193–196, 198, 205, 207
Heddon, Dee   29, 35
– *Walking Library*   29, 35, 37
Heidegger, Martin   95, 186
Hickson, Ella   3, 44, 88, 167
– *Oil*   44, 88, 167
– *The Writer*   3
Hill, Christopher   150
Hunt, Jeremy   140f.
Hunter, James Davison   168f.
Husserl, Edmund   188
Huxley, Aldous   12, 210
– *Brave New World*   12, 73, 210

Illouz, Eva   216
Ionesco, Eugène   106, 108f.

James, William   103f., 108, 110
Jameson, Fredric   4, 16f., 19f., 27, 30, 32, 98f.
Jean-Baptiste, Marianne   18f.
Jenkins, Henry   229
Johnson, Boris   89, 134, 136–140, 142, 159, 177
Jonson, Ben   202
– *The Devil Is an Ass*   202

Kalil, Hannah   87
– *Interference*   6, 87f., 90, 95–99
– *Metaverse*   87, 97–99
Kane, Sarah   20, 61, 152
– *Blasted*   61, 152
Kelly, Dennis   166
– "Dystopia"   166
Kermode, Frank   27, 29, 38
Kierkegaard, Søren   186

Index of Names

King, Dawn   88
– *Foxfinder*   88
Kirkwood, Lucy   5, 57, 61f., 64–66, 69, 88, 167
– *The Children*   62, 88, 167
– *Tinderbox*   5, 57, 62, 64–67, 167

Leadsom, Andrea   138f.
Lovecraft, H. P.   201, 204f.

Macaulay, Thomas Babington   150
Macdonald, James   20
Macmillan, Duncan   44, 167
– *Lungs*   44, 167
Malcolm, Morgan Lloyd   69, 140
– *Emilia*   69
Mallarmé, Stephane   105
Malm, Andreas   28, 31f.
Marcuse, Herbert   90, 93f.
Marlowe, Christopher   65, 202
– *Doctor Faustus*   202
May, Theresa   134–137, 139–142
McDonagh, Martin   65, 206
– *In Bruges*   206f.
McDowall, Alistair   7f., 185, 189, 192, 198, 204
– *Pomona*   8, 204f., 207
– *X*   7f., 185, 188–192, 198
Merkel, Angela   4
Milton, John   202
More, Thomas   133
Morris, William   153f., 157
– *News from Nowhere*   154
Mouffe, Chantal   5, 91
Moylan, Tom   2, 4, 57–59, 67f., 152, 160, 209f., 218, 221
Mullarkey, Rory   7, 149, 152f., 158f.
– *The Wolf from the Door*   7, 149, 152–162
Muñoz, José Esteban   33, 35
Murray, Andrew   11, 20

Nachtwey, Oliver   221
Nairn, Tom   157f.
Ngai, Sianne   161
Nietzsche, Friedrich   90

Nono, Luigi   39
– *La lontananza nostalgica utopica futura*   39
Norris, Rufus   143, 162

Orton, Joe   65
– *Entertaining Mr Sloane*   65
Orwell, George   79, 154, 210
– *1984 / Nineteen Eighty-Four*   59f., 62, 73, 78, 154, 210
Out of the Woods   28, 31

Parkinson, Katherine   15
Pasolini, Pier Paolo   219
– *Oedipus*   219
Patel, Priti   136f.
Paz, Octavio   104
Pearson, Morna   87, 97f.
– *Darklands*   87, 95, 97–99
– *Interference*   6, 87f., 90, 95–99
Pinter, Harold   65
Pollard, Clare   167
– *The Weather*   167
Powell, Enoch   66, 138
Powers, Richard   37
– *The Overstory*   37f.

Raab, Dominic   137
Raddatz, Frank   36
Rebellato, Dan   151, 153, 155, 160f., 198
Reckwitz, Andreas   168
Rees-Mogg, Jacob   139f., 159
Rimbaud, Arthur   105
Ronder, Tonya   167
– *Fuck the Polar Bears*   167

Sargent, Lyman Tower   57–59, 79, 209, 217
Sassen, Saskia   5, 44, 49f., 53
Scarry, Elaine   187
Scorsese, Martin   205
– *Taxi Driver*   205
Shakespeare, William   229–240
– *As You Like It*   235f., 240
– *Hamlet*   228, 231f., 236
– *Julius Caesar*   229, 234
– *King Lear*   236
– *Romeo and Juliet*   231, 236f.

– *The Merchant of Venice* 239
– *The Tempest* 235, 239
Sheldon, Joss 176
– *Individutopia* 176
Shelley, Mary 66
Sierz, Aleks 151f., 167, 176, 179, 189, 204f.
Skinner, Penelope 88
Sloterdijk, Peter 107
Smith, Stef 88, 167
– *Human Animals* 88, 167
Solnit, Rebecca 4f., 30f.
Stafford-Clark, Max 150
Stephens, Simon 8, 203
– *Pornography* 8, 203, 207
Stiegler, Bernard 32f.
Stock, Kathleen 180
Stoppard, Tom 59
– *Jumpers* 59
– *The Coast of Utopia* 59
Storey, John 133, 143, 154f., 160
Streeck, Wolfgang 6, 117–129
Swale, Jessica 235
– *Nell Gwynne* 235

Tannock, Stuart 133–135, 142
Terry, Michelle 229, 231f., 238f.
Thatcher, Margaret 4, 66, 124, 138f., 158
Theater des Anthropozän / Theatre of the Anthropocene 29, 36
– *Requiem for a Forest* 36

Thorne, Jack 88
Thorpe, Chris 6, 88, 90–92, 94f.
– *Confirmation* 91
– *Victory Condition* 6, 88, 90–92, 94f.
Tomlin, Liz 5, 43f., 46–54, 91, 160f.
– *The Cassandra Commission* 5, 43–45, 47, 50, 52, 54
Trump, Donald 43, 89, 124, 133
tucker green, debbie 2, 12, 17–19, 24, 61
– *hang* 2, 12, 17–19, 21, 24
Tundale 201f.
– *The Vision of Tundale* 201
Turkle, Sherry 95

Urry, John 226

Wade, Laura 2, 12, 14–16, 21f., 24, 151, 159
– *Britain Isn't Eating* 2, 12, 14, 16, 24
– *Posh* 151, 159, 162
Waters, Steve 167f.
– *The Contingency Plan* 167f.
Williams, Raymond 3, 94, 96, 202, 221
Wilson, Robert 29, 35f.
– *The Forest* 29, 35f.
Wittgenstein, Ludwig 106
Woynarski, Lisa 29, 37
– *The Celebrated Trees of Nashville* 29, 37

Žižek, Slavoj 160

Subject Index

Absurd / absurdity   111, 113, 153, 161, 176, 191, 197, 211, 220
– Absurdism   7, 112, 167
  – Absurdist dystopia   60, 167
Accessibility   8, 16
Affect   2, 5, 8, 13 f., 16, 21–25, 43 f., 47, 53, 61, 68 f., 88, 135, 190, 226
Affluent society   120
Agency   4, 6, 29 f., 32, 37, 45, 78, 87, 95, 97, 119 f., 124, 126, 129, 133, 152, 155, 158, 161, 217–220
Alienation   14, 93, 95, 215, 217
– Alienation effect   150, 160 f.
Allegory   49 f., 192, 204, 207
Anger   5, 15, 18, 34 f., 55, 69, 74, 77, 111, 135, 195
Anthropocene   29 f., 36, 39, 47, 166
Anthropogenic climate change   165–167, 170 f., 174, 177
Anti-capitalism   58
Anti-democratic   12, 88 f.
Anti-racist   19, 236, 239
Anti-utopia   58
Anxiety   1, 7 f., 10, 13 f., 21, 23, 60, 80, 96, 166 f., 185–198, 234, 236, 238
Apocalypse / apocalyptic   1 f., 5, 21, 27 f., 30 f., 33, 46 f., 50, 62, 78, 88, 92, 109 f., 153, 155, 161, 165, 170, 174, 179, 198, 205, 208
Aristocracy   155, 158
Artificial intelligence / AI   87, 97–99
ATTAC   121
Audience   2–11, 13 f., 16 f., 19 f., 22 f., 30, 35, 44–47, 49–52, 61, 68 f., 74, 78, 82 f., 91 f., 135, 139, 150, 152, 160 f., 172, 174 f., 179 f., 186, 192, 195, 203, 205, 218 f., 225–235, 237 f., 240
– Audience participation   8, 227, 231, 235
Austerity   117 f., 133, 135, 151
Authenticity   111, 168, 207
Authoritarian neoliberalism   6, 89
Avant-garde   6, 104–110, 114

Big Brother   158
Black Lives Matter   12, 14, 19, 74, 142
Body   15, 34, 53 f., 76 f., 80–82, 84, 92, 152 f., 166, 187, 194, 198, 206
Border   18 f., 137 f., 142, 185–187, 193–195, 198
– Border crossing   185, 194, 198
Boredom   190
Boundary   30, 193, 195
– Boundary crossing   185, 188, 193, 198
Brexit   19, 43, 124, 133–143, 151, 159, 162, 166, 168, 171, 185

Capitalism   4–6, 13, 16, 20–23, 28, 32, 49–52, 62–64, 66, 68, 94, 96, 98, 117–129, 150, 158, 166, 168, 173, 177, 180, 198, 205, 209, 216 f., 221
Capitalist realism   4, 156, 160, 176 f., 198
Capitalocene   28, 166
Catastrophe   20–22, 31, 38, 45, 54, 78, 88, 129, 134, 138, 166 f., 171, 173, 185, 190
Centre-left   120, 122 f., 126–128
Chronology   7, 112
Circuit of culture   217
City   3, 29, 37, 48–53, 82, 94, 99, 112–114, 204
Class   1, 4, 13, 73, 75, 103, 121 f., 124, 150 f., 157, 169, 176 f.
Cli-fi   6, 167
Climate   2 f., 28–30, 32, 37, 43, 61, 165, 167 f., 170 f., 174, 176, 178 f.
Climate catastrophe   5, 52, 87, 165, 176, 179
Climate change   5, 28, 34, 43, 46, 64, 165–171, 173–175, 177–180, 192
– Climate change theatre   5, 167 f., 173
Climate crisis   35 f., 43–47, 50, 88
Climate emergency   28 f., 32, 43
Collage   113
Comedy   7, 160 f., 168, 179
Comic   17, 22, 153, 156, 160 f., 195, 206
Commodity culture   93

Subject Index

Communication   31, 121, 189, 191, 193, 195–198
Community   2, 5, 8, 13 f., 16, 31, 36, 43, 48, 78, 81, 129, 133, 141, 150, 159, 162, 226, 228–230, 232, 235 f., 238, 240
Complicity   2, 5, 46, 48, 55, 68 f.
Conservative   66 f., 137, 140, 159, 180
– Conservative Party   64, 135, 139
Consumerism   155, 196, 198, 213, 215, 220
Consumption   212 f., 216 f.
Counter-hegemonic / counterhegemonic   2, 12 f., 24 f., 127, 129
Counter-narrative   143, 150, 210
Counter-site   74, 215
COVID / COVID-19   1, 4, 6, 8, 10, 16, 28, 30, 37, 50 f., 53 f., 87, 110, 134, 138, 141, 165 f., 171, 185, 225, 228, 230, 232, 234, 240
Crisis   4, 8, 11, 27 f., 31 f., 43, 45, 47, 60, 62, 77, 90, 113 f., 121, 124–126, 128, 165–167, 198, 207, 211, 218, 225, 229, 235, 239
Critical utopia   6, 57–59
Culture war   7, 168 f., 171 f., 177, 179 f.

Dadaists   6, 106 f.
Danger   28, 31, 88, 129, 157, 177, 194, 198, 210
Deep England   159
Deep time   30, 35 f.
Defamiliarisation / defamiliarization   76, 160, 218
Democracy / democratic   12, 14, 89–91, 117, 119–122, 124 f., 127 f., 136–140, 143, 150, 153, 162, 177, 180, 226, 240
Demon   8, 193, 205–208
Despair   4, 35, 60, 64, 190, 195
Destruction   5 f., 12–14, 20 f., 25, 27 f., 30, 37 f., 57, 66, 69, 83, 94, 103–108, 114, 129, 159, 185, 188 f., 191, 193, 195, 197 f.
Devolution   151
Digital theatre / digital stage   8, 36, 51, 110, 228 f., 232 f.
Disaster   5, 21, 28 f., 31–33, 54, 88, 93, 165 f., 168
Disaster utopia   30 f.

Discourse   2, 5–7, 43, 46 f., 49, 51, 61, 73, 99, 106, 108, 118 f., 134, 151, 154, 168 f., 171–174, 176 f., 179 f., 185, 187, 236 f.
Disorientation   21, 31, 105, 189, 213 f., 217
Disruption   6, 24, 33 f., 39, 43, 74, 78, 84, 106, 123, 133, 188–190
Dissolution   7 f., 188, 193, 198, 214–216, 218, 220
Divide / division   7, 15 f., 64, 133, 141, 166, 171, 178, 191
– Divided nation   149, 151, 162
Dungeons & Dragons   205
Dysfunction / dysfunctional   78, 189 f., 195 f., 198, 217, 221
Dystopian turn   58 f., 96, 167

Ecocriticism   6, 44, 47
Eco-dramaturgy   36
Eco-dystopia   167
Ecology   28 f., 35, 166
Eco-play   167, 174
écriture automatique   113
Elite   125, 150 f., 153, 157 f., 160, 162
Emotion   2 f., 8, 15, 31, 35, 47, 54, 79, 91, 94, 98, 103, 111, 134, 136, 165 f., 187, 189–193, 213 f., 218, 227, 229, 239
Empathy   34, 91, 98, 212, 217, 230, 235, 238
Empire   53, 69, 99, 105, 137, 139, 141–143
Empowerment   30 f., 78, 158
Engagement   5, 17, 47 f., 91, 161, 227 f., 233, 236
English Civil War   149, 151, 156, 158
Englishness   159
English Revolution   155
Environment   5, 28, 35, 37, 45, 49 f., 66, 52, 62, 79, 88, 98, 109, 126, 128, 170, 190, 198
Epic Theatre   34
Eschatology / *eschaton*   27, 109, 206
Establishment   139 f., 149, 156, 174
Estrangement   2, 24, 105, 209
Ethical witnessing   4, 29, 32–35, 39
EU / European Union   118, 121, 124, 126, 134–137, 139–141, 143, 159
Eutopia   5, 58 f., 69, 74

Evil 8, 103 f., 106–108, 110, 155, 201, 203, 205
Exclusion 84, 168, 180, 198
Experiencescape 226, 233, 240
Extinction Rebellion 29, 34

Fear 2, 7, 11–14, 22, 43, 62, 76, 78–80, 82 f., 92, 107, 165, 167, 185–188, 190, 192–195, 198, 206, 232, 235
Feeling 2, 6 f., 11 f., 15–18, 22–25, 34 f., 37, 44, 61, 76–78, 99, 107, 135, 143, 155, 186 f., 191–195, 212, 221, 233 f., 239
Feminism / feminist 5, 12 f., 17, 20, 24, 57 f., 60 f., 65–69, 177
Feminist dystopia 57, 61, 66–68
Financial crisis 11, 135, 207
Forest dramaturgy 29, 35
Fragmentation 14, 16, 95, 109–113, 125, 150, 169, 180
Future / futurity 1–9, 12, 14, 17–20, 22–24, 30 f., 33, 36–39, 45–49, 57, 59 f., 62–64, 73, 81, 87 f., 90, 95–98, 104, 119 f., 124, 133 f., 136, 138 f., 141, 143, 149, 152–155, 159 f., 166 f., 169, 172, 174, 185–188, 190–192, 194, 197 f., 211, 228, 233

Game logic 95
Gender 12, 57, 61 f., 67, 69, 151, 170, 180, 192
Global heating 185
Globalisation / globalization 53, 121, 126, 129
Globe Theatre 225–240
Glorious Revolution 159
Golden Age 7, 120, 133 f., 139, 143
Government 15 f., 82, 117 f., 120–122, 124–126, 128 f., 140, 149, 152, 158, 165, 174, 180

Happiness 12, 19, 21, 23, 103, 154, 211, 214–217
Hegemony / hegemonic 11 f., 24, 84, 119, 129, 150, 152, 158, 162, 172, 175, 178, 180, 210, 213, 221
Hell 4, 7 f., 20, 22, 31, 33 f., 96, 201–208
Heterotopia 74, 77, 79 f., 82 f., 84, 215, 220
Hierarchy 59, 198
Hope 2–6, 8, 15 f., 22 f., 27, 29–35, 39, 45–47, 58 f., 94, 112, 118, 129, 160, 179, 189 f., 218, 221, 229, 239
Humour 45, 160 f.
Hypercritique 32 f.
Hyperreal 216

Identity 14, 62, 81, 84, 91, 113, 133 f., 138 f., 143, 162, 166, 168–170, 172, 176, 180, 189, 198, 212, 217, 220, 238
Ideology / ideological 32, 66, 75, 83, 91, 95, 125, 127, 149, 150, 154 f., 160, 172, 180, 216
Imaginary 4, 46 f., 68, 129, 141
Imagination 5, 14–16, 19, 32 f., 35, 45, 54, 60, 62, 67, 73, 83, 88, 98, 134, 201, 216, 226 f., 233, 236, 240
Immediacy 1, 8, 209, 221
Inclusion 168, 180, 228, 239
Incongruity 156, 160 f.
Individual 8, 13–15, 21, 37, 49 f., 53 f., 77, 95, 104, 112, 117, 120, 123 f., 129, 158, 167–169, 174, 176, 194, 203, 210 f., 213–215, 217–221
Inequality 11 f., 14, 16 f., 23, 84, 123, 136, 160
Injustice 12, 18 f., 23, 28, 55, 84, 96, 109, 160
Insecurity 2, 13 f., 21, 194, 206, 238
Institutional racism 151
Irony 7, 57, 153, 157, 160 f., 202
Isolation 31, 93, 97, 99, 189–193, 198, 217, 225, 235 f., 238 f.

James Bond 159
Joint Stock 150
Justice 4, 6, 17–19, 128, 172

katabasis 202, 204
Keynesianism 117, 121

Labour Party 118–119, 122, 127, 142, 172
Language 2, 7, 13, 15, 17, 24, 33, 66, 105–107, 110 f., 113 f., 133 f., 139, 152, 165,

168f., 173, 175, 177, 179, 185–193, 195, 197f., 203, 212, 237, 239
Laughter   15, 160f., 174, 179
Leave campaign   134–137, 140f.
Left-wing   6, 118f., 125–127
Liberalism / liberal   16, 89, 91, 117, 120, 127, 141, 157
Lockdown   51, 87, 110, 141, 225, 228–230, 232–236, 238–240
London Riots   151, 157

Macbeth: A Conjuring   230
Manipulation   14, 16, 110, 193
Marginalisation / marginalization   74–76, 79, 154f., 161
Marxism   118, 149f.
Materialism   213
Metadrama / metatheatre   214, 218f.
Metaphor   53, 64, 95, 137, 202–204, 206
#Me Too movement   12, 238
Middle class   63, 153, 156f., 160f., 215, 225
Milibandisti   117, 119
Monarchy   149–151, 153, 155, 157f., 161f.
Morality Play   206
Multimediality   7, 186
Multiple casting   150

Nation / national   15f., 88, 120f., 125–129, 134, 138, 140–143, 150–152, 157, 159, 162, 168f., 172, 225
National Theatre   151, 167f.
National Theatre of Scotland   87
Naturalisation   75, 78, 83, 169, 177
Near future / near-future   68f., 87f., 123, 149
Near-future dystopia   3, 17, 60, 64, 90, 167, 185, 188
Negative utopia   17–19, 59
Negativity   6
Neoliberalism   2–8, 12–16, 19–21, 23–25, 50f., 88–91, 94f., 96f., 117f., 120, 122f., 127f., 129, 151, 155f., 157f., 160f., 167–169, 171–173, 176f., 179f., 209f., 214f., 217f., 221, 226
New Drama   60
New Labour   118, 125, 127

New Left   58, 118f.
New Writing   88, 96, 205, 235
Nostalgia   4, 6f., 39, 57, 122, 125, 127, 133–136, 138, 140–143, 154, 158f., 166, 190, 208, 232–234
– Nostalgiascape   226f., 232

Occupy movement   157
Online performance   *see* Digital theatre
Opacity   94, 192
Open Court   51
Ortonesque   65, 67

Pain   7f., 18, 34, 77f., 88, 98, 111, 126, 185–188, 192–198, 214
Pandemic   1f., 8, 13f., 28, 50–53, 104, 133f., 141, 165f., 185, 221, 226, 228, 230, 232, 237–239
Past   2, 4f., 7, 18f., 24, 33, 36, 39, 43, 46, 51f., 55, 83, 99, 108, 117, 127, 133–135, 137–143, 149, 153–155, 158f., 166, 185, 188, 190–192, 198, 227f., 233f., 238
Pathos   134f., 165
Patriarchy   13, 66f.
Performative space / theatre space   8, 35, 150, 226f.
Poetic language   24, 195
Polarisation   165f., 168–171, 177, 179
Political left   6, 119
Political theatre   2, 4f., 57, 59, 68
Populism   125f.
Post-apocalypse / post-apocalyptic   1f., 50, 78, 88
Post-Brechtian comedy   160
Postdramatic theatre   34
Postfeminist   61
Post-human / posthuman   29, 35, 38, 98
Post-pandemic   52
Poverty   13–16, 25, 118, 156
Power   18f., 49f., 53f., 59, 68f., 74, 76–78, 80–84, 87, 90, 98, 106, 118–122, 126, 128, 136f., 151, 153–157, 161f., 166, 168, 171, 180, 198, 202, 206, 213, 219, 221, 232
Pseudo dystopia   218
psychomachia   203, 206f.
Purgatory   206f.

Subject Index — 253

Race   13, 18f., 24, 74–76, 98, 141, 143, 170, 237f.
Racism   12, 18f., 75f., 84, 161, 166
Raiders of the Lost Ark   205
Realism   6, 32, 43, 50, 96, 153, 160
Regime   67, 92f., 142, 155, 157f., 166, 178
Repressive desublimation   93
Resistance   3, 30, 34, 68f., 84, 128, 151f., 154f., 161, 210, 218
Retrotopia   7, 154, 158f.
Revolution / revolutionary   4, 12, 17, 19, 33f., 59, 69, 93, 105, 149f., 152–162, 213
Rhetoric   7, 15, 21, 31, 68, 98, 119, 133–143, 165f., 168f., 171–173, 177
Right-wing   15, 89, 123, 125–127
Romantic / Romanticism   6, 46, 55, 95, 104f., 114, 216
– Romanticised   166, 170, 179
Royal Court   14, 17, 20, 51, 60, 62, 90, 161, 167f., 205
Ruin   6, 30, 51, 88–90, 94, 96f., 106, 193

Satire / satirical   3, 7, 15, 67f., 97, 158, 160f., 168, 170, 174, 176, 179, 205, 211, 214, 219
Second World War   107, 140f., 143, 159
Social democracy   58, 117, 120, 127f.
Socialist utopia   158
Space   2f., 3, 5–8, 14f., 18, 24, 27, 32, 35, 43, 45f., 49–54, 63, 67, 73–84, 91, 93–95, 97–99, 108–110, 112, 114, 120, 138, 150, 171, 189, 192, 209, 215, 217, 220, 228, 230–234, 239
Spatial order   73, 75f., 84
State-of-the-nation play   151
Stereotype / stereotyping   15, 168f., 171, 176f., 212f.
Structure of feeling   96, 221
Subject / subjectivity   68, 73, 77, 89, 91, 93f., 133, 161, 168, 217–221
Supremacy   18f., 25, 91

Surreal / surrealist   6, 21, 113
Symbolism   6, 104f., 114

Technology   6, 14, 87, 90, 92, 95–99, 136, 139, 166, 170, 177, 185, 190, 193
Temporality   2, 7, 18, 24, 52, 63, 88, 94f., 186, 189f., 192f.
Terror   13, 35, 60, 76f., 80
Terrorism   13f., 137, 208
Thatcherism   118, 151
Theatricality   29, 78
theatron   2, 4, 8
theatrum mundi   36, 109, 201
Third Way   127, 172
Threat   2, 13f., 27, 81, 89, 98, 107, 137f., 143, 151, 167, 175, 177f., 180, 185, 198, 205, 217, 220f.
Tory   139
Trope   8, 46, 49, 88, 143, 167, 178, 201, 221

Ukania   157–159, 161
Unhappy archives   2, 12f., 16, 24
Upper class   73, 155, 159
Upper-middle class   156
Urban landscape   37, 44
Utopian performativity / utopian performative   2, 4, 22f., 35, 39

Verbatim   91, 143, 151
Violence   6, 69, 83, 105, 114, 153, 156, 159f., 161, 195, 198, 209
Viral dystopia   53
Viral theatre   11
Virtual   8, 87, 97, 240

Whig   150, 157
Working class   22, 120, 124f., 127, 135, 137, 225

Young Adult Dystopia   73
Young Adult Fiction   3

www.ingramcontent.com/pod-product-compliance
Lightning Source LLC
Chambersburg PA
CBHW020227170426
43201CB00007B/340